THIS RADICAL LAND

THIS RADICAL LAND

A Natural History of American Dissent

DAEGAN MILLER

The University of Chicago Press
Chicago and London

The University of Chicago Press, Chicago 60637
The University of Chicago Press, Ltd., London
© 2018 by The University of Chicago
All rights reserved. No part of this book may be used or reproduced in any manner whatsoever without written permission, except in the case of brief quotations in critical articles and reviews. For more information, contact the University of Chicago Press, 1427 E. 60th St., Chicago, IL 60637.
Published 2018
Printed in the United States of America

27 26 25 24 23 22 21 20 19 18   1 2 3 4 5

ISBN-13: 978-0-226-33614-5 (cloth)
ISBN-13: 978-0-226-33631-2 (e-book)
DOI: 10.7208/chicago/9780226336312.001.0001

Library of Congress Cataloging-in-Publication Data

Names: Miller, Daegan, author.
Title: This radical land : a natural history of American dissent / Daegan Miller.
Description: Chicago ; London : The University of Chicago Press, 2018. | Includes bibliographical references and index.
Identifiers: LCCN 2017042893 | ISBN 9780226336145 (cloth : alk. paper) | ISBN 9780226336312 (e-book)
Subjects: LCSH: Nature—Effect of human beings on—United States. | Thoreau, Henry David, 1817–1862. | Environmentalism—United States.
Classification: LCC GE150 .M55 2018 | DDC 304.20973—dc23
LC record available at https://lccn.loc.gov/2017042893

♾ This paper meets the requirements of ANSI/NISO z39.48-1992 (Permanence of Paper).

For all the wild ones

The trees alongside the fence
bear fruit, the limbs and leaves speeches
to you and me. They promise to give the world
back to itself. The apple apologizes
for those whose hearts bear too much zest
for heaven, the pomegranate
for the change that did not come
soon enough. Every seed is a heart, every heart
a minefield, and the bees and butterflies
swarm the flowers on its grave.
The thorn bushes instruct us
to tell our sons and daughters
who carry sticks and stones
       to mend their ways.

HAYAN CHARARA, "Elegy with Apples, Pomegranates, Bees, Butterflies, Thorn Bushes, Oak, Pine, Warblers, Crows, Ants, and Worms"

CONTENTS

When the Bough Breaks   1

ACT ONE At the Boundary with Henry David Thoreau   15
ACT TWO The Geography of Grace: Home in the Great Northern Wilderness   47

Intermission   97

ACT THREE Revelator's Progress: Sun Pictures of the Thousand-Mile Tree   105
ACT FOUR Possession in the Land of Sequoyah, General Sherman, and Karl Marx   161

Enduring Obligations   213

Acknowledgments   229
Abbreviations Used in Notes   235
Notes   239
Index   313

WHEN THE BOUGH BREAKS

Then the coal company came with the world's largest shovel,
And they tortured the timber, and stripped all the land.
Well, they dug for their coal till the land was forsaken,
And they wrote it all down as the progress of man.

JOHN PRINE, "Paradise"[1]

What happens when the past's oldest witness comes crashing down dead?

A new day will dawn . . . but over what?

Where are we, who are we, when the bough breaks?

Bostonians opened their eyes on a Wednesday morning in 1876 to opaque February skies bleakly blanketing a city made suddenly strange. At 7 p.m. the evening before, the enormous Great Elm on the famous Boston Common had been toppled by a hard wind.[2] Of course, trees fall all the time with never a thought spared them, but the Great Elm was different. It was famous in the nineteenth-century as an emissary from the past, and it appears ubiquitously in prose, poem, and print, a people's treasured heirloom, believed to be among the last living witnesses to the young nation's milestones; its loss was disorienting. Paul Revere, on his 1775 midnight ride, was rumored to have passed by the tree.

Puritans supposedly strung up witches and dissenters to dangle from its branches. And the tree was there during what many white settlers imagined to be the final days of Massachusetts's coastal American Indians, bearing mute testimony to the beginnings of what would someday come to be called Manifest Destiny.[3] The Great Elm had been cherished as the city's teller of these stories that no one alive had experienced; and then one morning in 1876—the centennial year—it was gone.

Americans had long been used to learning their history from trees, and they gave to them a name: witness trees. The term was originally used by surveyors, who, as they went about their business of dividing one plot of earth from another, chopped a mark into long-lived bark so as to make the invisible lines of property legible. Because trees can live to be hundreds of years old, each witness tree came to be an archive of knowledge about who belonged where long after generations of surveyors had been laid to rest. Even after the term wandered from the surveying profession into popular culture, it retained its core connotation—a witness tree was a tree that had once seen something remarkable, a repository for the secrets of the past.

The Great Elm, an American elm, was a witness tree, though it had never been a boundary marker; its fate would be something much grander. It was, instead, used to stake a claim to the entire Boston area, and ultimately to New England, by rooting a particular people in a particular landscape—it is conspicuously there, the lone tree on the Common, just to the left of the powder and watch houses, on the earliest surviving map of the city, John Bonner's 1722 *The Town of Boston*, a map obsessively reprinted and distributed throughout the nineteenth century.[4] It's easy to overlook at first, and yet it also stands out, for the tree is the single still moment in a map that is otherwise all transient motion—arterial roads and converging ships—a map celebrating modern mobility and the circulation of merchant capital. A map, in other words, tipsily weaving between Progress and the past, movement and quietude, financial markets and nature.

Progress. By the middle of the nineteenth century, the word was on everyone's lips, an elegant one-word distillation of one of the time's most commonly heard phrases, "westward the course of empire takes

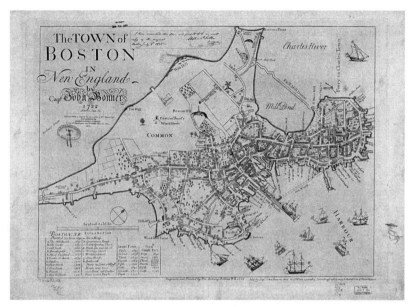

FIGURE 1. John Bonner, *The Town of Boston in New England* (Boston: Fra. Dewing, 1722), Library of Congress.

its way."[5] It was an era when the improved steam engines of capital were revving hotter and faster, annihilating space and time, sending the juggernaut of modernity rocketing into the future, all to the thrill of many Americans. Alexis de Tocqueville, the astute French social critic, caught the spirit of the time during his tour of the United States in the early 1830s: trying to determine why America had no poets worth the name, he argued that their eyes had been "filled with another spectacle." "The American people sees itself advance across the wilderness," Tocqueville continued, "draining swamps, straightening rivers, peopling the solitude, and subduing nature"—though many Americans would have protested that Tocqueville had missed the point.[6] Paper and pen might have been all right for antiquated Europeans whose story had long been written, but modern American poetry was written on the land, using the surveyor's transit and the pioneer's ax. The United States, by the mid-nineteenth century, was no longer an exemplary city upon a hill but a mature nation seizing its golden destiny and aided by the precocious children of modernity—the quick-as-lightning telegraph and the unstoppably bullish steam engine—bringing the distant near, conquering nature's chaos, and chasing away once and for all the dark European aristocratic past with the beacon of democracy.

What made this drama seem peculiarly American to commentators on either side of the Atlantic was the starring role of nature. Tocqueville understood it as a wilderness Americans needed to break in their great leap forward, but many in the United States saw the landscape as a thing to perfect: if America was to be the new Eden, then the unruly wilderness had to be "improved." Americans were simply finishing work that a Christian God hadn't quite gotten around to. Why else would Providence provide limitless space and a bottomless reservoir of natural resources? This was the creation myth of a nation that, above all else, believed in the moral imperative of limitless growth.[7] And so the mid-nineteenth-century was the "age of go-ahead": the age of capital, an age in which the narrative elements of free space, free resources, free markets, and an exceptional, individualistic, rationally self-interested bootstrap-pulling-up people became the key tenets of the nation's official faith in Progress, a trinitarian belief in nature, nation, and capitalism whose good book was the land itself.[8]

This all seemed like the birth of something completely new, and the character of American Progress gained an influential spokesman in 1836, when Ralph Waldo Emerson's *Nature* broke upon the literary scene. "Our age is retrospective," it famously begins: "The foregoing generations beheld God and nature face to face; we, through their eyes. Why should not we also enjoy an original relation to the universe. . . . Why should we grope among the dry bones of the past?" A new nature is, in the end, on the New World's side, Emerson argued, and Americans needed to pay attention to their landscape if the United States was to realize its potential: "Nature, in its ministry to man, is not only the material, but also the process and the result. All the parts incessantly work into each other's hands for the profit of man."[9] Even advanced technology is but "reproductions or new combinations" of natural phenomena. There are all sorts of ways that one could read passages like these (and it is always dicey to pin Emerson down to one interpretation), although the essential point is Emerson's supreme confidence that nature and the nation will emancipate each other, ushering in a new golden age. "Know then that the world exists for you," he concluded. "For you is the phenomenon perfect."[10]

Yet, the easy celebration of Progress was often tempered by a nagging sense of disconnection and growing powerlessness: if everything is always in motion, then how can we ever put down roots? Such anxiety is there, in the guise of ambivalence, in Bonner's map, which, alongside motion, bespeaks a clear desire for the rootedness symbolized by the elm. Indeed, according to the map's legend, the city wasn't founded in 1630; it was planted, and Boston was a city as natural as the American elm. What better way for nineteenth-century Bostonians to reassure themselves that they truly belonged in a land that had only recently been home to a vibrant American Indian culture than to stake a claim to the ancient past with a witness tree? After all, one of the things trees do is hold a shifting world together.

In Nature's Nation trees starred in the stories Americans told themselves about their place in the world, and this was partly because in this epic whose text was the landscape, the most prominent feature was the continent's leafy verdure. At the moment of European contact, something like 45 percent—or 850 million acres—of the future contiguous United States was forested, and of that, more than 680 million acres were located east of the Great Plains—the region soon to contain the bulk of the nation's population.[11] An ocean of trees rolled in an unbroken wave from Maine to eastern Texas, New England to the Deep South, until it broke upon the grassy plains. But even when they were absent, trees were present: the most distinguishing feature of "the vast, barren, and treeless" Great Plains was the region's lack of trees.[12] Trees also stood out because they were the longest living things a person could encounter. And so while trees are always nonhuman beings "out there," in the nineteenth-century United States, they were also always profoundly human metaphors that helped Americans make their world make sense.[13]

But a metaphor, without its surrounding sentence, is merely a lonely word, and a word alone means next to nothing. As with words, witness trees made sense only because they remained rooted in particular landscapes, the terrestrial equivalent of a sentence, a paragraph, a book. And like books, landscapes are a mix of the human-made and the natural. Look closely at any landscape that you know well—the backyard with its grass and trees and landscaping; the neighborhood whose poverty was long ago underwritten by marshy ground and frequent flooding; the farmed field whose glacial soil is suitable for growing grass but not wheat; the forest scene with its logging trails—and you'll see a world partly physical but also cultural. Or perhaps you are more at home indoors than out and know landscapes best as paintings and photographs, for landscapes need not be physical; they can also exist as maps and essays, poems and travel narratives and music. A landscape, unlike an environment, with its strong scientific connotation of the objectively material and scientific, is always a profoundly human creation made out of profoundly nonhuman stuff, and while environments can exist independently of humans, landscapes vanish the moment humans disappear from the scene.[14] As Henry David Thoreau put it in *Walden; or, Life in the Woods* (1854): "Wherever I sat, there I might live, and the landscape radiated from me accordingly."[15]

This means that we can double Thoreau's sentence back upon itself; we can begin with the landscape and follow it backward to where a person lived, where he sat thinking; we can follow the landscape back into a person's mind and watch as she dreams it into being. Every landscape is ultimately symbolic, its outward appearance a jumbled record of particular human ways of living in and making the world, its shards refracting particular ideas about who and what belonged. All landscapes, in other words, are histories, and as histories they can be read, if one can find an introduction. In each of the histories that I recover in this book, a witness tree stands as the central, organizing, orienting landmark that points the way to a forgotten past. Each essay depends upon a tree (or a forest of them), and what unfolds beneath its branches is not the typical tale of American progress, but something more radical: histories of dissent, of freedom, of equality, and of justice. These are the subjects of this book.

There is, for instance, *Lessons from an Apple Tree* (1848), a children's story, which begins, innocuously enough, by saying that nothing is more common, very early in the spring, than to see a pair of red-breasted robins busily building a wattle-and-daub nest for their sky-blue eggs in the fork of a tree. So common that perhaps the nesting goes unnoticed by most. "But," the author writes, "let us just *consider* a little of what the tree, the nest, and the birds may suggest":

> All things and all creatures are bound together, and live and flourish together. You may see the apple-tree as it supports the nest, tied, as it were, to the sun by rays of light, opening the little mouths in its leaves and roots to drink in the falling showers, bathing its branches in the all-surrounding air, and getting new breath and new life every moment; you may see this in your thoughts, and a great deal more, and see only what is strictly true. Now as it is with the apple-tree, so it is with you. You cannot exist for an instant without help coming from abroad; you are joined to the creation, and to your fellow beings; you do not stand alone, and cannot live alone.
>
> Looking at this truth, what does the apple-tree and the bird's nest say to selfishness; to boys and girls who take every thing, use every thing, enjoy every thing as if it were their own, made only for them, and never dream that they have any return to make for all the goodness which cares for them, or that they have any thing to do for others?[16]

Something started to shift in America during the 1840s, a decade of especially belligerent landscape rhetoric. Democrats in the 1844 presidential campaign who fantasized about the United States seizing the Oregon Country all the way to Alaska's southern tip adopted as their slogan "Fifty-four forty or fight!" A year later, the United States annexed Texas, and a year after that invaded Mexico, thus snatching up the rest of the Southwest and California. At the same time, the phrase "Manifest Destiny" exploded into print, finally giving a name to the sensibility Tocqueville had observed ten years earlier.[17] The idea that God himself had given his imprimatur to America's continental pretensions made the flexing of national muscle exhilarating for many; yet others worried that a devilish American military was making the nation safe not for freedom, but for slavery. This was no idle fear, for the years immediately following the Mexican-American War saw the rise of the "Second Middle Passage," the era when slavery quickly swarmed west into Texas from its traditional stronghold in the Upper and Lower South, threatening Kansas and Nebraska, looking to clear land and settle wherever a slave could make his owner a dollar.[18] By the middle of the 1840s the peculiar institution seemed to be the economic lifeblood of the entire nation.

At the same time, the United States was industrializing, and factories were colonizing the banks of rivers throughout the northeastern part of the nation, appropriating water that farmers had long monopolized. These factories required a workforce of willing laborers—many with farming roots recently severed by industry—tethered to their machines, producing commodities, and thus profits, for an absent boss who profited the most from their hard work. Of course, the raw material that fed many of those factories—cotton—was watered with the blood and sweat of African American slaves. Yet the slave economy was never only a way of organizing society; it was also a spatial regime that brutalized the land.[19] And unfortunately for the nation's witness trees, the last half of the nineteenth century saw industrialized logging so voracious that many feared the country might run out of timber altogether.[20]

Behind all of this profit-driven violence lay one of the foundational intellectual transformations of the modern world: alienation. It's difficult to define precisely what modernity is or when it began, but most scholars seem to agree that the modern world is sustained by (at least) two main intellectual taproots that took hold of Western European thought sometime around the sixteenth century: a notion of time that broke sharply at the present and a startlingly new conception of nature.[21] Both depended on the estrangement of alienation, on wrenching apart things that had once seemed to cohere. Consider time. Ever since the nineteenth century, to be "modern" has meant to be new, and newness implies that time is not cyclical, but linear; it can be broken into past, present, and future, or the old, the new, and the newest. Along with linearity came the notion of Progress: things inevitably get better as an enlightened today turns into tomorrow and leaves yesterday in the dark ages of the past. Suddenly, too, was the natural bounded off from the human. No longer was the world and everything in it governed by an unknowable living spirit, but by mechanical, rational laws; and no longer were humans part of nature, but unique beings with reasoning minds capable of making their own fate by discovering and mastering the laws of the universe. One of the great dangers of alienation is that it seems, inevitably, to lead to hierarchy, and just as the golden present represented progress over the darkness-enshrouded past, so too was rational human consciousness celebrated as history's ultimate achievement, while an erstwhile living world was drained of its vitality and reconceived as an inert, well-oiled machine whose inner workings could be exactly divined, controlled, and exploited by a rational, skeptical, scientific human mind. And with this, the intellectual revolution was complete—humans were no longer a part of anything, beholden to nothing but their own future.

Such thinking was supposed to set us free, supposed to break the chains of tradition, religion, and nature, ushering in a golden age of human Progress. But by the middle of the nineteenth century it was also starting to become clear that modernity brought with it unavoidable and crushing violence. The problem was this: an alienated nature became a collection of natural resources, good only for its human use. And those alienated humans who had the bad fortune not to have descended from a few supposedly enlightened, light-skinned Western Europeans also became resources, *human* resources—the field or factory "hands," the "labor force"—in the drama of Progress directed by, and for, the elite. Modernity and capitalism are not synonymous, though they are often used that way; but modern thinking did provide the intellectual conditions for reenvisioning land and labor as merely raw resources in the production of capital, and capitalism is the economic, social, and cultural regime that has come to dominate the modern world.

Rather than an Emersonian emancipation born of the immaculate conception of nature, nation, and capitalism, the course of American empire, of the budding capitalist economy, even of Progress itself, seemed, to many critics, most clearly marked by a bruising legacy of domination—of landscapes blighted and lives ruined for the privilege of a very few.[22] Perhaps modernity was not a rebirth, but the ascendance of a world "radiant with triumphant calamity."[23] Emerson himself worried as much. In 1846, ten years after *Nature*, his *Poems* debuted, and with them one simply entitled "Ode." It's a cutting poem that decries the "famous States / Harrying Mexico / With rifle and knife!" and "the jackals of the negro holder"; it rails against "He who exterminates / Races by stronger races / Black by white faces," and then elevates its rhetoric to condemn a growing antihumanism in American culture:

> Things are in the saddle,
> And ride mankind.
> There are two laws discrete,
> Not reconciled—
> Law for man, and law for thing;
> The last builds town and fleet,
> But it runs wild,
> And doth the man unking.[24]

Something was shifting in the 1840s, and when the apple tree gave its lesson in 1848, only a year before the gold rush that would see thousands from all over the world flock to California in a cutthroat battle for easy riches that decimated American Indians, blasted through hillsides, and choked streams with muddy tailings, there is something radical in its insistence on community—a community that includes "all things and all creatures"—something radical in the question "what does the apple-tree and bird's nest say to selfishness?" It is an unresolved question that shivers its unsettling way well beyond the antebellum United States into our today.[25]

There were dozens of alternative currents pulsing their way through the nineteenth century: Sabbatarians who wanted to worship on Saturday and Spiritualists who spoke to the dead every day of the week, free lovers and no lovers and Graham-cracker-munching denouncers of masturbation, enlightenment-seeking hash eaters and those who damned demon rum—and not all of them grounded their vision of the good life in the land. Nor did all reform currents look ahead to a more just future: many critics of American Progress longed for a supposedly simpler era of authentic premodern craftsmen, aristocrats, saints, and noble savages, times when everyone knew their place and stuck to it. Historians have tended to see ambivalence toward the modern emphasis on getting ahead as ironically contributing to the cultural hegemony of capitalism by failing to look soberly into the haggard face of the modern world and ask, "How did we get here, and what can be done?"[26] There were also those, like the antebellum slavery apologist George Fitzhugh, who dismissed modern notions of freedom and equality outright as culturally toxic; for them, only a return to feudalism would ensure a kind of progress.[27]

But rejecting modernity was only one of the many possible routes of resistance; for some the goal was not flight into a soft-focus past—a conservative, therapeutic reaction—but the creation of a fulfilling, just, green present. Why should progress not give all of us more comfortable, more fulfilling lives? Why should progress not leave all landscapes greener, cleaner, healthier, more beautiful? What good were steam engines and cheap cotton cloth if dull landscapes of degradation, slavery, exploitation, and monopolization were the costs? In the modern world, objects may lack vitality, and resources—whether natural or human—may have no value other than for commerce; yet there were those, call them countermoderns, who refused to let the injustice of objectification go uncontested. For these reformers and radicals the point was to reanimate what modernity had drained of life. The point was to reenchant the world, for an enchanted world is one that demands something of its inhabitants, if only the realization that the boundary between "us" and "them," "human" and "nature," is, like all boundaries, imaginary.[28] And what better way to come under the spell of countermodernity than to listen to the stories told by trees? After all, the word *radical* is related to the word *root*.[29]

The American religion of Progress has left many stories inscribed in a long series of deep scars across the landscape. But a counterhistory of patient, vital resistance has long worked those jagged outlines into something softer.[30] Delicate blades of green things grow, as they always have and always will, crumbling concrete and shifting foundations, causing microscopic cracks to bloom into vast furrows, letting in the light by which new landscapes appear—sunshine revelations of possibility. Some of these green living things grew into witness trees that were taken to tell deeply ambivalent tales about the course of empire, an ambivalence springing from the word *witness* itself: a noun and a verb, both transitive and intransitive; a thing testifying to a past and prophesying a future; a survivor of crime, as in Emily Dickinson's 1858 poem—

> I Robbed the Woods—
> The trusting Woods.
> The unsuspecting Trees
> Brought out their Burs and mosses
> My fantasy to please.
> I scanned their trinkets curious—
> I grasped—I bore away—
> What will the Hemlock—
> What will the Oak tree say?[31]

But also a lone figured bathed in the heavenly headlight of divine revelation, a pathetic fallacy impossible to believe (can trees really speak?) and the supreme legal arbiter of territorial truth.

The testimonies of these witnesses were used to actively contest the gale of Manifest Destiny that seemed to blow forever westward, and to redefine what modern America stood for; they sought to redefine the very notion of progress itself.[32] One way to do so was by rooting culture in place, which was, in many cases, an act of dissent: this was, after all, the age that celebrated the annihilation of space as one of the keys to empire. And so certain resilient witness trees became central landmarks tangling place and culture together, under whose spreading branches alternative landscape visions—alternative moral narratives about who and what belonged in the world—came alive.

This book is my attempt, in four essays, to listen to those witness trees, to read what is written on their leaves, to get down on paper the lessons those once-living metaphors might still hold for today. The book explores four different nineteenth-century countermodern landscapes, each set in a different part of the United States, each embodying a different kind of modern American progress, each united in a shared conviction that the world is alive, that our world should be better. The essays range from the 1840s to the close of the century, and their starring human characters are many. Some, like Henry David Thoreau, are well known; others, like the abolitionist James McCune Smith, the photographer A. J. Russell, and the anarchist Burnette Haskell are less so. A red, Californian Karl Marx will make an appearance, though not until after we chart a river in Massachusetts whose name means "harmony," brave the unicorns guarding the way to Timbuctoo in New York State, and stand squinting into the future, alongside the tracks of the transcontinental railroad, in a dry Utahan breeze.

With the right eyes, you'll be able to see that, for all its distance, the dehumanizing, alienated world of the nineteenth century seems oddly familiar to those of us who today are threatened by the climate change of runaway global capitalism, by the violence perpetrated daily against the poor and people of color, by the great industrially driven die-off that Elizabeth Kolbert calls the sixth great extinction, by the cancer-causing toxins that buoy Fortune 500 profits—and we are all, every one of us, threatened.[33] We who are now alive have inherited this world from those who came before. But because Americans have been wrestling with the environmental, cultural, and social costs of modernity for as long as the idea of America has existed, we who now live in hot times have also inherited a long, though often overlooked, legacy of vitality and enchantment, a twinned legacy of social justice and environmental sensitivity, a legacy preserved by witness trees.

* * *

When the green countermodern players that inhabit each essay cast their eyes over their landscapes, they saw a once-forested land stubbled with broken boughs. But each was also a dreamer, looking for meaning in the testimony of a tree. Though the world might have appeared shattered, they focused not on the splinters, but on the sunlight streaming between, which carried with it the dim outlines of something better, greener, more just—a land of healthy trees and living rivers, a commonwealth of communities and farm fields and humane technology, a vision of free nature and free humans all in mutual sympathy. They intuited that landscapes are always made, are always in a moment of becoming, and that critique is meaningless unless it changes the world.[34]

ACT ONE AT THE BOUNDARY WITH
HENRY DAVID THOREAU

We wonder whether the dream of American liberty
Was two hundred years of pine and hardwood
And three generations of the grass

And the generations are up: the years over

We don't know

ARCHIBALD MACLEISH, *Land of the Free*[1]

1

A steel-guitar breeze, shimmying through willow's whips, makes them sigh, involuntary and vital. But for the accidental planting of a willow's fertile catkin in the mud at water's edge, the dominant sound on this stretch of the Concord River wouldn't have been sylvan, but a round, liquid purling mixed with the labored rumbling of wagons, carts, and coaches on the two straddling roads to the south and west.[2] A willow loves water, and when its embryonic seed came to rest on the Concord's bank, an improbably hair-thin root called the radicle emerged first, and burrowed downward, against the day, seeking the river, finally to mingle its rooted permanence with the water's ever-changing flow.

Aboveground, the tree's fingerlike leaves yearned for the morning sun even as the willow cascaded down over the bank, providing shade, the illusion of permanence, strings for the wind to sound.

When Henry David Thoreau stood under this tree behind his dear friend William Ellery Channing's house, in the summer of 1859, readying his boat for the day's river adventure, the wind would blow; and when it did, he heard a harp, or sometimes a violin—an impossible, and therefore sublime sound, one he groped for the right words to describe, finding them only in repetition and revision: "It told me by the faintest imaginable strain, it told me by the finest strain that a human ear can hear . . . that there were higher, infinitely higher, planes of life which it behooved me never to forget."[3] Or maybe he heard nothing at all. Maybe he was too busy pitching notebooks and field equipment over his boat's gunnels and chatting with Channing about which channel of the river they would spend their day observing. Maybe the willow went unheard, but it wouldn't have gone unnoticed: Thoreau often shot his gaze toward the significant notch he had cut low in the tree's trunk earlier that summer.[4]

Thoreau stood on a river's bank in the midsummer of 1859 contemplating a ragged wooden gash because he was being paid to survey the river, and that notch was the key to gauging the river's height. Although Thoreau is now remembered mostly as a romantic nature writer, in his own time and place he was a highly trained, well-regarded, disciplined though eccentric land surveyor.[5] It was a job he often loved. He was good at it—he had the right temperament—and he had been hired to survey the river because it flooded.[6] In twenty-two miles the Concord fell only thirty-two inches—it was very nearly a pond—and any additional water heaved the river up and over its banks before gravity's current slowly siphoned it out to sea. This was a good thing, and the annual springtime deluge was the town's lifeblood, because the flood always rolled back, leaving behind it a thick, black nutrient-rich muck spread all across the bottomlands, whose field grasses grew fat and sleek on nature's bounty, perfect fodder for the farming town's livestock.[7]

But in 1798, in the predawn haze of the industrial era, a century-long fight for the Concord began when the Middlesex Canal Corporation downstream at Billerica increased the height of an old mill dam that had been flung across the river in 1704. Later upgraded yet again in 1851 to service a mill, this improved Billerica dam jacked the water's level higher through Concord, impeding the river's already-sclerotic return, keeping the waters unnaturally high well into the summer, ruining the hay, threatening the town's livelihood, and setting in motion a wave of lawsuits and petitions to the state legislature, which always seemed to be resolved in the interest of the mills and their capitalist owners, lasting throughout the nineteenth century.[8]

At the root of each dispute over the river's water was an argument over the meaning of *improvement* and its near cousin, *progress*. Improvement could mean the cultivation and refinement of human qualities, but it also had a material connotation: that to improve the earth was to mix one's own labor with it, to cultivate untouched, wild nature into an Edenic garden, and thereby to claim it as one's own. This was the preferred usage of the "improvers," those farmers who advocated for a scientific approach to husbandry that emphasized conservation, personal rectitude, and practices akin to what we might now call permaculture. There were improvers throughout Concord, and they even had an organization, the Concord Farmers' Club.[9]

Yet, the word *improve* was diverging from its agrarian roots, developing into something with the flavor that the word currently holds: new, improved technological, economic, advancement.[10] And Concord's improvers found themselves surrounded on all sides by these capital-minded lovers of Progress, because Thoreau's corner of Massachusetts was also ground zero for American industrial capitalism and its textile-mill-based Industrial Revolution: the American factory system had gotten its start with the famous Lowell mills in the early 1820s, on the Merrimack River, just upstream from the spot where the smaller Concord joined its faster-flowing cousin.

By 1859, the dam at Billerica had passed into the hands of a textile-factory owner named Charles Talbot, and soon after, upstream farmers noticed that floods flooded deeper and longer than ever. However much Talbot's dam checked the flow of the already-languid river, its effect was magnified many times over as dozens of mills periodically released their water into the river, in effect flooding Concord from both ends. Finally, the growing city of Boston needed clean water for its teeming, increasingly industrialized masses, and that water had to come from somewhere—which turned out to be the overworked Concord watershed. To compensate country folks for the water diverted toward the city, the Boston Water Board constructed a number of ponds up and down the Concord's spine whose timed releases would keep the Concord's levels uniformly high, even during periods of drought.[11] Concord's farmers were predictably incensed by all of this and convinced that the culture of industry conspired to swamp their livelihoods: "These lands," began one familiar with the debate, "are the most valuable in the State, for farming purposes, there is no doubt. . . . [I]t seems too bad that they should be rendered almost worthless, merely to accommodate a few old mills."[12]

And so a riddle threaded its way through Concord: what is the best use of a river? To power mills that would churn out cheap commodities, aid the growth of towns, yield fantastic profits for their owners, and help usher in an age of modern industrial capitalism; or, as Concord's farmers argued, to grow a productive, profitable, improved farming landscape?[13] It was just such a question that Thoreau, who stood straining to catch the quick words whispered by a willowy witness, was hired to answer.

2

Thousands of trees similar to Thoreau's willow have dotted the American landscape inscribed by thousands of ax-wielding men. From the moment of European footfall, surveyors have been cutting marks into trees to lash the chaotic, polyphonic story of the landscape into one coherent narrative, just as Thoreau cut his mark into the willow to hear the Concord speak and so solve its riddle; Thoreau and his tree were both inheritors of a much longer history whose deep context was, in Thoreau's own day, spreading its way across the continent. As America steamed west on rails of iron, there was always a surveyor in the vanguard, and the pantheon of American heroes includes many a man who navigated his way into the history books with map and compass: Daniel Boone, Lewis and Clark—even George Washington, Thomas Jefferson, and Abraham Lincoln spent time with surveying's tools.

Though blazing a tree with an ax was as old as surveying itself, the connotation of every mark began to change with Thomas Jefferson's Ordinance of 1785, a law that proved foundational to the growth of the infant nation. In the colonial era, land had been settled haphazardly, but Jefferson sought to bring order to territorial expansion through a streamlined, efficient process of rationalized surveying and cartography. A new nation, in the New World, needed a method of self-invention. That's why Americans, beginning with Jefferson, inscribed the national grid—which now stretches from the streets of Manhattan, across the checkerboard of the Great Plains, all the way to the beaches of California—onto the land.[14] Anchoring this imaginary, cross-hatched latticework to the earth were the witness trees standing sentinel at the corner of many six-mile-square townships, America's building block.

Today, the grid might seem so obvious as to appear natural, but it was truly revolutionary for its time, and it caused a stir upon its introduction to American soil in the late 1670s when William Penn brought the idea to his mid-Atlantic colony. Penn and his surveyor general, Captain Thomas Holme, dreamed of a city—Philadelphia—with orderly, well-mannered streets converging always at right angles. Their shared vision was a response to the twin scourges of plague and widespread fire—the wages of ignorant city planning—that swept through London in 1665 and 1666. But the City of Brotherly Love was to be modern and rational, immune to pestilence and safe from inferno, both banished by the almost magical power of the intersecting line. The design was daring in its unnaturalness, yet it seemed to work, and by the 1780s, would-be town forefathers across the new nation were modeling each of their own little cities upon a hill on Penn's innovation.[15]

But it wasn't until after independence, when Jefferson latched onto the idea as a way to ensure orderly continental settlement, that the grid took on truly national proportions. The grid's greatest appeal was that it was powerfully abstracting: it transmuted real, impossibly complicated land into simple squares on a map, all of which looked the same; this abstraction allowed for clarity, for a welcome relief from the headaches, political intrigue, and disenfranchisement that seemed to come with the traditional system of settlement and surveying that had long characterized Jefferson's South—a free-for-all of indiscriminate corruption by "surveyors" whose only claim to qualification was their claim to being qualified. In the colonial era, it wasn't at all uncommon for the same exact parcel of land to be sold to many different buyers at the same time, and legal wrangling over who owned what clogged Southern courts for decades.[16] Jefferson's plan, therefore, was to map the land *before* settlement, to pinpoint its resources and divide the nation's territory into uniform squares, then to parcel them out intelligently to deserving yeoman farmers.[17] And, like Penn before him, Jefferson was guided by faith:

> Those who labour in the earth are the chosen people of God, if ever he had a chosen people, whose breasts he has made peculiar deposit for substantial and genuine virtue. It is the focus in which he keeps alive that sacred fire, which otherwise might escape from the face of the earth. Corruption of morals in the mass of cultivators is a phaenomenon of which no age nor nation has furnished an example. . . . It is the manners and spirit of a people which preserve a republic in vigour. A degeneracy in these is a canker which soon eats to the heart of its laws and constitution.[18]

The Republic's very soil was composed of democracy, Jefferson writes, and in those who tilled it and planted it, who fed its transubstantiated body to their children, who rooted themselves in it, in these people lay the hope of the nation and of the world. "We have an immensity of land courting the industry of the husbandman," he wrote—this was what made America exceptional. But without the grid to ensure fair and equal distribution, none of Jefferson's democratic, agrarian utopia would come to pass, and the sylvan "manners and spirit" of the people would sour and rot from the twin cankers of urbanization: landlessness and wage dependency. It was the grid—clear and rational and simple—that would show the country the way back to Eden.[19]

By the mid-nineteenth century, surveying—and its first cousin, exploration—was the stuff of breathless excitement, and one can feel the thrill of discovery breeze through the era's popular culture. It's there in J. N. Reynolds's 1839 story, *Mocha Dick; or, The White Whale of the Pacific*, an account of Reynolds's sailing expedition to Antarctica, during which he heard tales of a great white whale that rammed whaling ships, a tale which Herman Melville later spun into his massive *Moby-Dick; or, The Whale* (1851).[20] When Darwin's *The Voyage of the Beagle* debuted in 1845, it became the "Victorian equivalent of a bestseller," and the daring surveyor-explorers, especially those who headed expensive, ambitious, multiyear projects that scoured the American West, were counted among the leading men of their age: the John C. Frémonts, F. V. Haydens, John Wesley Powells, Clarence Kings, and George Wheelers.[21] They were the romantic geographic surveyors of the 1840s to 1870s who crossed deserts, starved on mountain peaks, plumbed the Grand Canyon, and their reports often read like adventure tales of strangeness, hardship, and sublime beauty.[22] It was into this cultural current that Thoreau launched himself, just twenty-four years old and looking for a vocation, in 1840.

3

Thoreau loved the Concord River, and he would come to pursue the discipline of surveying with an obsessed intensity in part because it allowed him to learn the river's daily rhythms, its hardly noticed secrets. The river was wild and ultimately unknowable in its ever-changing habits, and yet surveying, Thoreau was convinced, would help him fix its living likeness. By the 1840s, Thoreau was exploring surveying's farthest and most arcane corners—the unsolved mysteries of terrestrial magnetism; the discipline's remote history; the best, most finicky methods of ensuring a compass's accuracy—and he quickly became a passionate scholar of surveying whose research went far beyond his daily tasks of measuring property boundaries and town lines, laying out woodlots, and planning roads.[23]

It's not entirely clear when Thoreau learned the art of surveying, or why—although surveying was often taught as part of a mathematics course, and when Thoreau and his older brother, John, opened a school in 1838, it was one of the subjects they covered.[24] But he did learn, and though he had some help from one of the town's older surveyors, Cyrus Hubbard (who periodically lent the upstart tools as well as, one imagines, trade advice), Thoreau was at root a bookworm. So he did what any bookish person does—he went to the bookstore. One day, he came home with Charles Davies's *Elements of Surveying*.[25]

Like many surveying manuals from the period, Davies's *Elements* rhythmically drummed into its reader the never-varying importance of exactitude, a cadence that guided Thoreau's footsteps in the field. Davies issued strict commandments about the importance of ascertaining precise declination—the ever-changing difference between true north and magnetic north—and the proper habits of taking a sighting, both rituals that Thoreau observed strictly, in all kinds of weather, during all times of the day, always wary lest inattention corrupt the precision of his instrument.[26] The authority of the compass was of such importance to Thoreau that by 1851 the language of surveying bled evermore into his prose, even into one of his most famous essays, "Walking," which retains the earthy scent of the field: "When I go out of the house for a walk," he wrote, "I . . . inevitably settle southwest. . . . My needle is slow to settle,—varies a few degrees, and does not always point due southwest, it is true, and it has good authority for its variation, but it always settles between west and south-west."[27]

This obsession with accuracy and variation earned Thoreau his reputation as one of Concord's premier surveyors, and it was this reputation that got him his job surveying the Concord River, in 1859. He had been hired on June 4 by Simon Brown, an elite, improvement-minded farmer and the chairman of the Committee of the Proprietors of the Sudbury and Concord River Meadows, an organization that arose both in response to Talbot's grass-killing dam and in hopes that the State of Massachusetts's newly appointed Joint Special Committee, which had been formed to look into the flooding issue, would finally rule in the farmers' favor.[28] Brown hired Thoreau to measure and catalog the width of each bridge that crossed the Concord between Sudbury and Billerica, the characteristics of every pier that might jut its obstructing pylons into the water, the history of each bridge's construction and improvement, and the character of the falls at Billerica itself. Doing so, Brown's committee hoped, would help prove that "the great oppression and spoliation to which we and our fathers have been so long subjected" was the fault of industry (and more specifically, the fault of Talbot's dam at Billerica), not sluggish nature.[29]

Between June 22 and August 22 or thereabouts, Thoreau spent thirty-four days on the river, often with his friend Channing, sounding the bottom in hundreds of places—stopping every thousand feet from Sudbury to Billerica, a distance of more than twenty-five miles—and seeing with his mind's eye the river bottom from the side, seeing its gullies and hills.[30] He also plunged into the town's archives, searching out maps he should consult and local citizens whose long memories he could mine for crucial information.[31] He took thirty-three pages of detailed notes, boiled them down to one single 15" × 32½" document, and finally submitted his chart, which exactly fulfilled every instruction, to the committee, whose members must have been pleased with Thoreau's work.[32] Charles Davies would have been proud, but Emerson was in a huff over what he saw as wasted time spent away from his neighbor's *true* work—writing: "Henry T. occupies himself with the history of the river, measures it, weighs it, strains it through a colander to all eternity," the elder transcendentalist had written to a mutual friend.[33]

Emerson was right to worry. Thoreau was zealously disciplined in his surveying, which isn't surprising, because most nineteenth-century surveying manuals, including Thoreau's copy of Davies's *Elements*, demanded exacting precision, not only in one's instruments, but also in one's body and mind. Surveying was as much a way to measure the land as it was a method of plotting the perfect, modern man: the would-be surveyor needed to be polished and upright; he needed "a careful training of the eye and hand . . . the exercise of judgment in laying out and prosecuting the work . . . physical alacrity and endurance in the collection of the data. . . . In short, the requirements are such that only a person of good judgment, temperate habits, active temperament, and scholarly attainments can hope to excel in this profession," as Lewis Haupt's surveying manual from 1883 put it.[34]

4

Despite Jefferson's best intentions, the grid—and with it, Western surveying and cartography in general—has come to be vilified as abstracting, disconnecting, even antidemocratic in its enabling of capital. Take abstraction: it's true that the grid translated local, particular places into abstract spaces. This was the entire point, for the attraction of abstraction lies in its ability to shear off the distracting particularities that give anything its complicated character, and thus bring clarity to an overdrawn picture. Complications can be fatal; we abstract to avoid them. We abstract to focus on generalities. We abstract to connect—land is land; forty square acres anywhere is equivalent to forty square acres everywhere else.

But if abstraction brings clarity and connection to disparate things flung across great distances, it can do so only through severing the tangled roots that plot any thing in a local place. Abstraction is a way of saying that some general connections matter and that others, the more specific ones, don't. What must be ignored in order to think that a parcel of land in one place is like a similarly sized parcel anywhere else? One way to answer that question is to look at what was abstracted out of the picture—people: squatters, wanderers, and, above all else, American Indians; and also topography, flora and fauna, and ecology. The grid turned real places into abstract spaces, and it was that perfect grid of mitotic squares, spreading out effortlessly into the sunset, that helped eventually give Americans the idea that the continent was empty, the land free for the taking as long as a well-trained property surveyor was near at hand.[35] Jefferson had hoped that his abstraction would roll democracy and his chosen people of God into one tight landscape of liberty, but he hadn't adequately reckoned with the metastasizing forces of capital, and it didn't take long for the economic elite to grasp that Jefferson's grid could be folded into a funnel to efficiently siphon off the wealth of an entire continent. All landscapes serve somebody's purpose.

Once abstraction standardized the landscape into generic, lifeless parcels, and then imaginatively emptied it of all prior claims, the profit hungry could reimagine land as yet another fungible product to be traded on the open market, like any other commodity. This is the point of commoditization, a special kind of abstraction: every sack of grain, every board foot of lumber, every gallon of oil, every good, is stripped of its individuality and imagined to be exactly like every other commodity of its class. The value of everything is then connected to the market, is totted up against the dollar, which opens the entire world to trade—that's the genius of capitalism.[36] Between the days of Jefferson and those of Thoreau it became increasingly possible to shift one's attention from the content of the landscape to its borders, from the productive value of the land to its exchange value as a unit. It became possible, between Jefferson's time and Thoreau's, to see land as essentially like a dollar. And like a dollar, land was neatly divisible—if you couldn't afford to purchase a whole 640-acre section, you could break it for change: half sections, quarter sections, and on down to the ubiquitous 40-acre plot, the terrestrial equivalent of a penny.[37]

Of course, it's too simple to say that the older communal land ethic died away instantly and without a trace. Indeed, much of the tradition of resistance to capitalism that we have in the United States begins with the assumption that land should benefit all of us (or at least more of us), just as American environmentalism is premised on the notion that some land is unique. Such values have been written in the ground, all across the nation: one can take refuge from the gridded bustle of Fifth Avenue in New York City's delightfully circuitous Central Park or, if hankering for something wilder, wade into the woods at one of the nation's national parks. Or one can settle for something closer to home and more humble, like the centrally located public green spaces—a descendent of the commons—that lie near the hearts of many Midwestern towns. And the garden city movement of the later nineteenth century, with its lawns and curvaceous, tree-lined streets, was an attempt to plan urban spaces for human, rather than economic, needs.[38]

Nevertheless, the grid paved the way for the freightliner of American capitalism, and the nineteenth century did see the steam-powered advance of a marketplace whose very existence depended in part on the abstraction, disconnection, and commoditization of land.[39] For instance, Chicago's volcanic rise to world economic prominence in only a few decades was made possible by a few enterprising land boosters and the gridded, "empty" space that they sold and resold for ever-increasing profits. Chicago's Board of Trade, founded in 1848, existed for no other reason than to profit from the commoditization of grain, which was grown on the vast squares of its commoditized hinterland stretching to the city's south and west. "The striking and peculiar characteristic of American society," observed the French political scientist Émile Boutmy, in 1891, "is that it is not so much a democracy as a huge commercial company for the discovery, cultivation, and capitalization of its enormous territory."[40]

A final fact: the very first big-business trust in America, from the late eighteenth century, was a real-estate outfit named the North American Land Company.[41]

5

How do you describe your love?

This was a question that Thoreau struggled his entire life to answer.

The first step, of course, is to figure out what lodestone it is that your eye and mind and body all swing toward.

For Thoreau, this was easy: along with its eponymous river, Thoreau loved the woods of Concord.

But then what: do you gaze furtively, or explore with eye and hand? Do you listen, anticipate, converse?

In the fall of 1851, Thoreau resurveyed the town's line. It was an old New England tradition for the Concord selectmen to walk the boundaries and reacquaint themselves with the town's sylvan witnesses once a year. But because of legal issues that had arisen, this was the first time that the selectmen had chosen to include a surveyor in their number.

And that surveyor was Thoreau.[42] Though he didn't let on to it, he was excited, even honored, to be asked, and he put the adventure down twice in his journals: once in the surveying notebook that he had been keeping since 1849, using disciplined surveyor's prose—"perambulated the line between Concord and Acton, from a split stone near Paul Dudleys"—but also a second time in his personal journal:

> Commenced perambulating the town bounds . . . Mr. —— told a story of his wife walking in the fields somewhere, and, to keep the rain off, throwing her gown over her head and holding it in her mouth, and so being poisoned about her mouth from the skirts of her dress having come in contact with poisonous plants . . . —— described the wall about or at Forest Hills Cemetery in Roxbury as being made of stones upon which they were careful to preserve the moss, so that it cannot be distinguished from a very old wall.
>
> Found one intermediate bound-stone near the powder-mill drying-house on the bank of the river. The worker-men there wore shoes without iron tacks. He said that the kernel-house was the most dangerous, the drying-house next, the press-house next. One of the powder-mill buildings in Concord? The potato vines and the beans which were still green are now blackened and flattened by the frost.[43]

A funny thing happened to Thoreau every time he looked at his compass: his mind strayed from its close focus on straight lines, and here he is, the trained, exact surveyor expertly wandering, running seemingly unconnected things—gossip, stone walls, surveying, labor exploitation, phenology—all together, creating a subjective, historical, experiential patchwork of a map at the very same time that he was employed as the legal guardian of the town's boundaries.[44] His entries tell us of death's economic balance sheet: powder leveled troops as well as mountains and forests—"The willow reach by Lee's Bridge has been stripped for powder," he had written a month before his perambulation.[45] They tell us of commerce and communication with the larger port city of Boston. We learn that poisonous plants grow, and we hear what might be a bit of loose talk, or possibly a morality tale about vanity, or maybe a lesson about the loss of local, rooted knowledge; we hear of early efforts at historic preservation. Thoreau tells us something about the attempts to mitigate the hazards of large-scale industrialization and the defense industry: workers in the powder mills—workers helping to make the munitions for the Mexican-American War—wore shoes without hobnails so that the simple human act of walking, of striking one's feet on the ground, wouldn't kick up sparks. And finally, Thoreau takes the time to notice the change of seasons, to notice that the window in which certain crops could thrive was shuttered tight against the cold.

His journals are filled with these moments. When he surveyed the Bedford town line in 1859, Thoreau noted boundaries, of course, but he also plotted the year's natural history: "I hear the *te-e-e* of a white-throat sparrow. I hear of phoebes,' robins,' and bluebirds,' [*sic*] nests and eggs. I have not heard any snipes boom for about a week, nor seen a tree sparrow *certainly* since April 30 (??), nor *F. hyemalis* for several days."[46] And on June 10 of that year, he mentioned that he "cut a line, and after measured it, in a thick wood which passed within two feet of a blue jay's nest, which was about four feet up a birch, beneath the leafy branches and quite exposed. The bird sat perfectly still . . . while we drove a stake close by."[47]

Thoreau learned to survey in order that he might discover how to get lost; to triangulate between nature and humankind; to pinpoint the connections that emplace a person in a deep context at once cultural, economic, and natural; to find himself. "The boundaries of the actual are no more fixed and rigid than the elasticity of our imaginations," Thoreau had written in 1853, and the deliberate, blending, countermodern confusion he sows between inanimate property and living creatures is a key to what Thoreau used his compass for.[48] As always, at the heart of his observations there is a sense of wonder: an acknowledgment that no matter how precise one's instrumentation, the well-trained mind will always stumble upon the unknown, the fantastic, the enchanted.[49]

Simon Brown and the other members of the Committee of the Proprietors of Sudbury and Concord River Meadows hired Thoreau because he was exact, and they hoped that Thoreau's skill would keep the river in their control.[50] They were right; they won. (At least temporarily. Although the state ordered the Billerica dam torn down, Talbot appealed, and when he did, the state changed its mind.)[51] But Thoreau took the job so that he could spend time with one of his muses. Whether Brown and the rest of the committee knew it or not, in Henry Thoreau they had chosen an observer for whom the Concord River was everything: "For the first time it occurred to me this afternoon what a piece of wonder a river is,—a huge volume of matter ceaselessly rolling through the fields and meadows of this substantial earth," he had written in one of his very first journal entries, in the spring of 1838, clearly in love.[52] By 1859, he knew that piece of wonder intimately: for nearly a decade he had been filling his journal with hundreds of entries on the river's height and temperature, both of which he would continue to monitor until, dying of tuberculosis that had been aggravated by a bitterly cold day spent counting the rings of his other love, the town's trees, he could no longer leave his bachelor's bed; he had skated most of the Concord's length in the wintertime, had floated on it with his brother John in 1839, had imagined it in his first book, *A Week on the Concord and Merrimack Rivers* (1849). He recorded the industry— the farming, the munitions makers, the mills—on the river's banks, the wildlife along its shores, as well as his own physical immersion in the landscape: "Bathing at Barrett's Bay, I find it to be composed in good part of sawdust, mixed with sand."[53]

And he did discover something new while he sweated through hot July days for his employers, something that hadn't, until that month, caught his eye during years of river gazing, something that made his labor well worth its price. Though he never did fully settle on whether he thought the flooding of Concord's meadows was anthropogenic or not, there was no doubt in Thoreau's mind that the river's flow was being artificially influenced.[54]

Back in August 1854, Thoreau had noticed that in periods of drought, the river remained fairly high.[55] Then, while surveying for the Committee of the Proprietors in the summer of 1859, his willow tree told him a curious thing: the river had risen when not a single drop of rain had fallen.[56] So, over the course of the summer, he took hundreds of readings of the river's height at various times every day, until he discovered that the river had an irregularly rhythmic pulse. And then he discovered that the pulse wasn't the river's at all.[57] It was the action of the upstream mills, pumping their water-powering floods during the workday and holding them back at night and over the weekend when no gears needed turning, that explained the ebb and flow.

The river was merely an organic machine, a river rationalized—gridded, even—to serve the interests of capital, just like the land, and Thoreau, miserable, wrote of a body "so completely emasculated and demoralized . . . that it is even made to observe the Christian Sabbath."[58] Even worse, the river was forced to tell factory time: "By a gauge set in the river I can tell about what time the millers on the stream and its tributaries go to work in the morning and leave off at night, and also can distinguish the Sundays, since it is the day on which the river does not rise, but falls. If I had lost the day of the week, I could recover it by a careful examination of the river. It lies by in the various mill-ponds on Sunday and keeps the Sabbath. What its *persuasion* is, is another question."[59]

But it's just here at the revelation of the river's regulation that Thoreau got sidetracked and wandered away from the desolate gloom of his statistics toward an uncaulked chink through which streamed hope. What *is* a river's persuasion? Thoreau knew, from years of painstaking measurement, that water will always force its own course, given the time, and that the path of least resistance sometimes leads directly through the most concrete redoubt of Progress: "Who knows what may avail a crow-bar against the Billerica dam," Thoreau had written back in 1849 in *A Week on the Concord and Merrimack River*, his book-length poem to his town's life-giving stream.[60] In 1856, he discovered that the sympathetic river had beaten him to the punch: on a visit to another dam, Thoreau found himself "amused with the various curves of water which leak through at different heights. . . . The dam leaked in a hundred places between and under the planks, and there were as many jets of various size and curve."[61]

"It excites me to see early in the spring that black artery leaping once more through the snow-clad town," wrote a Thoreau whose faith had taken a knock from his findings but had not, ultimately, succumbed to the logic of surveying. "All is tumult and life there, not to mention the rails and cranberries that are drifting in it. Where this artery is shallowest, *i.e.*, comes nearest to the surface and runs swiftest, there it shows itself soonest and may see its pulse beat. There are the wrists, temples of the earth. Where I feel its pulse with my eye. The living waters."[62] Despite the mills' regulating effect, the river was alive, untamed even if it had been calmed. Precision again led him to bewilderment, fact to a flowing unknown. "Men are inclined to be amphibious, to sympathize with fishes, now," he had written in 1852, in a joyful spirit of boundary crossing.[63]

The river was alive, and somewhere, from the branches of a nearby tree, the white-throated sparrow sang *te-e-e*.

6

The ultimate problem with abstraction is that it relies always on violence—one of the word's many meanings is "to steal," and one of the purposes of the nineteenth century's great federally funded land surveys was to prepare for stealing land from others.[64] The army had its own corps of engineers, which was reorganized in 1838 to better reconnoiter the continent's people—especially the American Indians—along with its exploitable natural resources; and John C. Frémont, the flamboyant, self-promoting explorer who would take the name "Pathfinder" in his run as a presidential candidate in 1856 ("Free Soil, Free Men, and Frémont" was his campaign slogan), was sent out on a series of secret missions between 1842 and 1845 to help make the developing narrative of violent annexation—of Oregon, California, Texas, and the southwestern lands taken after the Mexican-American War—appear inevitable, natural, divinely sanctioned.[65] The grid never picked a pocket. After the gun smoke cleared and the earth soaked in all the blood it could hold came the surveyors with their squares, rewriting history, declaring the land open, empty, and free for democracy.

Yet, rather than freedom, Jefferson's grid instead helped to knit slavery ever deeper into the nation's fabric. In those red-hot years of Manifest Destiny following the Mexican-American War, slavers and their peculiar notions of property streamed west to join the slaveholding pioneers who had long claimed the territory as their own. It was Anglo slaveholders living in Mexico's northern province who goaded Texas to revolution in 1835 after Mexico had outlawed slavery, and it was the prospect of nearly 270,000 square miles of slave-worked Lone Star land that helped push the Southern-dominated federal government toward provoking war with Mexico. Once the shooting was done, slave owners could enjoy an unbroken view from Florida all the way to the borders of free-state California, and what they saw was the promise of fortune.[66]

Though the grid laid the groundwork for American capitalism, it was the slaves' sweat and tears that transmuted raw earth into one of the most valuable commodities the world has ever known, the one thing that turned the United States from a backwater into a busily humming, modern economy—cotton. Cotton production exploded during Thoreau's lifetime, from eighty million four-hundred-pound bales in 1811 to well over a billion at the time of his death, and from 1840 on to the Civil War, cotton never fell below 50 percent of total US exports. All that cotton was grown on land abstracted away from others; all that cotton was picked by stolen labor—nearly a million slaves were marched west out of the Upper South between 1790 and 1860; and one of the most important parts of a cotton-picking slave, from an owner's perspective, was his or her hands, so highly prized that the part replaced the human whole in the slave owners' mind such that a living, breathing, dreaming, person became merely four fingers and a grasping thumb.[67] After the enslaved had picked and cleaned his master's cotton, it was shipped to textile mills, where the raw fibers were spun into cloth by another set of hands—these, usually white but also valued merely for their profit margin. Much of the slave-picked Southern cotton was ultimately destined for the factories pockmarking England, which, by 1860, was home to two-thirds of the world's textile industry.[68] But some of that cotton went to Massachusetts, where the Lowell mills alone consumed a hundred thousand days of enslaved people's labor every year, and some of that Massachusetts-bound white fleece found its way farther upstream, to a mill along the banks of the Concord River, whose dam backed the water up ever closer to a willow tree with a notch cut low in its trunk, under which stood a surveyor, thinking.[69]

When Thoreau looked down the river, he saw a current congealed by a factory in which tired humans toiled to make white cloth from slave-picked cotton grown on land won through violent dispossession. He saw a world that he loved collide with Progress. He saw a landscape ruled by the profit motive—"this world is a place of business," he elaborated in "Life without Principle." He continued, "I think there is nothing, not even crime, more opposed to poetry, to philosophy, ay, to life itself"—a landscape of violent abstraction, of gridded lives, land, and water, a place of thievery where everything complicated, unique, and wonderful was put up for sale.[70]

Thoreau's genius lay in seeing the invisible, countermodern connections between things; but what use is talent if it can't be shared? Though he was gifted with a rare ability to notice, his life's burden lay in trying to give his visions corporeal form, and so make them apparent to others. At this task, he struggled; there's no one place where Thoreau ever plainly plotted the intersection of land, labor, and capital, the meeting ground of Southern chattel slavery and the Northern wage system, the overlapping territory of machine, cotton, and surveying.

He could feel the connections, he knew they were there, although his writing largely remained archipelagic; and yet connections do exist in places like the famously incongruous first chapter of *Walden*—a book ostensibly about living in the woods—called "Economy," a blistering jeremiad hurled at those he would later mock as the "champions of civilization."[71] "I cannot believe that our factory system is the best mode by which men may get clothing," he wrote; "the principal object [of the factory] is, not that mankind may be well and honestly clad, but, unquestionably, that the corporations may be enriched."[72] Perhaps he was pondering the profit motive as he walked to town one July evening in 1846 to pick up a worn-out walking shoe—he had been living at Walden for over a year—when he was interrupted by Concord's constable and thrown into the town's three story, granite-blocked, double-grated-window jail for refusing to pay a mandatory tax that would have gone to support the then-raging Mexican-American War.[73] He spent only a single night in the lockup, but when the sun rose the next day, it dawned on the beginnings of Thoreau's great essay "Civil Disobedience" (1849) — the one that influenced Martin Luther King Jr. — an essay pointing out that, although it might be easy to hold Southerners solely accountable for slavery, an even greater burden of blame fell on "a hundred thousand merchants and farmers here [in Massachusetts], who are more interested in commerce and agriculture than they are in humanity."[74] Thoreau would return to the issue of slavery throughout the final decade of his life, in essays with names like "Slavery in Massachusetts" (1854), in lines like "perhaps olive is a fitter color than white for a man," and in three fiery pieces dedicated to John Brown, who sought, in 1859, to stage a slave rebellion by first capturing an arsenal in Harper's Ferry, Virginia: "A Plea for Captain John Brown" (1860), "Martyrdom of John Brown" (1859), and "The Last Days of John Brown" (1866).[75]

Nor was Thoreau alone in critiquing the problems of his age: the transcendentalists filling Concord and Boston with their meetings and pamphlets questioned nearly everything under the sun, from the abuse of prisoners to the role of women, from slavery — Thoreau's tax resistance was inspired by the anarchistic communard Bronson Alcott, who was arrested for the same abolitionist crime in 1843 — to workers' rights, from eating meat to competitive individualism.[76] Dissent could even be found in that most unlikely of places: mainstream politics. Beginning in the 1830s, the Jacksonian Democrats from the North increasingly found themselves at odds with those down in Dixie over the issue of slavery. Often thought of as the party of workers, small farmers, and a limited, centralized state opposed to internal improvements, the Bank of the United States, and growing corporate interest, the Democrats also had a reputation as Southerners. But some Northern Democrats came to question the right of one person to hold another in perpetual bondage, and they actively contested the role of the slave power in determining national policy. A few years later, the Republican Party sprang to life and committed itself to limiting the territorial expansion, though not the practice, of slavery. There were even radical land-reform advocates who made up part of the Northern Democratic left, folks like George Henry Evans, a freethinker, abolitionist, workers' advocate, and agitator for the free, democratic distribution of the nation's land to anybody suffering from poverty—including American Indians and African Americans.[77] Evans was associated with some of the best-known communitarians and utopian socialists of the day, including an abolitionist from upstate New York named Gerrit Smith who would, in 1846, deed 120,000 acres of Adirondack land to three thousand black New Yorkers in a bid to rid the United States of racism.

But what separated Thoreau from many of his politically radical peers was his insight that free trade and slavery, the mill and the factory, territorial expansion and offensive war and demoralized rivers, all were rooted in a peculiar kind of landscape. Thoreau had gone to Walden in 1845, the year *Manifest Destiny* entered the American vocabulary, a year when his surveying interests were growing into a passion, in order to "drive life into a corner, and reduce it to its lowest terms, and, if it proved to be mean, why then to get the whole and genuine meanness of it, and publish its meanness to the world."[78] He went searching for meanness, and of course found it, for if the culture of capitalism fed in part on reconceiving of nature as natural resources, it grew fat on reducing people to mere functions.[79] Yet life, as it must, endured. "I have thus surveyed the country on every side within a dozen miles of where I live," he wrote, and what he discovered was sublime: he found the wild.[80]

Unfortunately, there is perhaps no way Thoreau has been more misunderstood than as an advocate for humanless wilderness, and such misunderstandings often branch from this concept of his—the wild.[81] "In wildness is the preservation of the world," he famously wrote in his ever-unfinishable essay "Walking," a line that is *still* misquoted as a defense of untouched wilderness.[82] But wildness, a quality, and wilderness, a place, are not the same things.[83] Wildness gives life its character— "Life consists with wildness. The most alive is the wildest"—it is the underground current sustaining all nature, including all human nature.[84] Wildness is what pulsed through Thoreau and the leaping black artery of the Concord; wildness is why, in *Walden*, Thoreau was able to count mice, phoebes, ants, loons, a "winged cat," ducks, and dozens of animal others among his neighbors, and runaway slaves as his visitors.[85] "All good things are wild and free," he wrote in "Walking," an essay that also happens to seethe with scorn for capitalist surveying.[86]

If Thoreau thought that in wildness the world could be preserved, it was because the wild was a landscape where "no fugitive slave laws are passed," because the wild planted every human as "an inhabitant, or part and parcel of Nature," because the wild grows untamed and always threatens the abstract orderliness of the grid.[87] Thoreau had chased life into a corner, and when he did, he saw that the market's abstractions were neither natural nor inevitable, though his neighbors might live "lives of quiet desperation" believing that "there is no choice" besides the world of profit.[88] Yet, perhaps Thoreau boasted a bit much. Perhaps he didn't chase, but instead was led by the hand until, alone together, life turned to face its follower. When it did, Thoreau felt his own boundaries fade and float out on the day's light. Though desperation might march over the land in tidy squares, a wild nature paid no heed to imaginary straight lines traced across the sand. "The life in us is like the water in the river," he wrote in *Walden*'s conclusion. "It may rise this year higher than man has ever known it, and flood the parched uplands."[89]

7

What is the best use of a river? That's the question that Simon Brown and his committee had hired Thoreau to find out, and I'm sure they had little patience for wildness, or wandering, or border-bending metaphors, or any of the other transcendental things for which Thoreau used his surveying skills. Brown, like Concord's other farmers, wanted legally defensible maps and tables of statistics from his hired man. Thoreau knew this; he complied with his employers' wishes and was paid, and then he often found himself crossing into the monochromatic kingdom of despair. He knew the company that surveyors kept—he had written that the devil himself practiced the trade—and he fretted over the deadening effect that measurement played on his mind.[90] It was the surveyor's job to abstract, after all, to produce a map claiming to capture something essential, something *real* about the earth; but Thoreau worried that in gaining a critical distance he was losing the world.[91] He was in a particularly bad state in the thinly lit January days of 1858, when, after weeks spent sighting lines back and forth through the woods surrounding Walden Pond, he wrote:

> I have lately been surveying the Walden woods so extensively and minutely that I now see it mapped in my mind's eye—as, indeed, on paper—as so many men's wood-lots, and am aware when I walk there that I am at a given moment passing from such a one's wood-lot to another's. I fear this particular dry knowledge may affect my imagination and fancy, that it will not be easy to see so much wildness and native vigor there as formerly. No thicket will seem so unexplored now that I know that a stake and stones may be found in it.[92]

Though Thoreau loved the mysterious poetry of old maps, he also lived his life in a running battle with them, sniping at their inaccuracy, bemoaning their makers' overconfidence, worrying always over a map's stealthy power.[93] "How little there is on an ordinary map," Thoreau would write soon after finishing his work for Simon Brown. "How little, I mean that concerns the walker and lover of nature. Between those lines indicating roads is a plain blank space in the form of a square or triangle or polygon or segment of a circle, and there is naught to distinguish this from another area of similar size and form."[94] These blank spaces irritated Thoreau, because they weren't, in fact, blank, and his famously circuitous writing can be seen as a literary effort both to picture the landscape more truthfully and to confound a map's simpleminded linearity.[95] It's a bitter irony, then, that what Thoreau produced for the Committee of the Proprietors was the ultimate abstract landscape: a wilderness of statistics.

Thoreau wasn't asked to produce a map for the committee, because its members already had one, and he came to know it well.[96] It was originally drawn in 1834 by B. F. Perham under the supervision of the highly regarded Massachusetts surveyor Loammi Baldwin, and it was

FIGURE 2. Right half of Loammi Baldwin's Concord River map. Loammi Baldwin, *Plan of the Concord River from East Sudbury & Billerica Mills* (detail), 1860. Courtesy Concord Free Public Library.

still considered legally up to date in 1859. It's a map whose greatest strength is its generic beauty, a map that looks like any other, a map that outlines the river's arcing course, the area's political boundaries, the towns' notable buildings. Five bridges exist on the map, and at the very top, the river's most-watched statistics: length, 22 miles, 802 feet; vertical fall, 2.865 feet. It's a map confidently offering itself as objective, naturalistic, reliable, passive, a true transcript of raw, uninhabited space. All of it aggravated Thoreau.

Where are the people, he wondered? Where are the fields and the trees and the homes and the roads and the stores? Where are the landscape's other inhabitants—the plants, the animals, all the things every bit as real as the river and courthouse and political boundaries?[97] Where, he must have asked himself, am I?

On July 7, 1859, Thoreau was poring over the Baldwin-Perham map as part of his work for the committee, filling his journal with tables of measurements charting distances, and falls, and all kinds of statistical minutiae.[98] He listed all of the bridges, and then corrected the names that Baldwin and Perham got wrong.[99] Then, in early July, Thoreau began a map of his own because, at some point, he had become disillusioned, not only with Baldwin but also with the mute stupidity of his own statistical table. What could a column of numbers have to say about a living river that was worth hearing?

And so Thoreau improved the time: "I [am] reminded of the advantage of the poet, and philosopher, and naturalist, and whomsoever, of pursuing from time to time some other business than his chosen one,—seeing with the side of the eye. The poet will so get visions which no deliberate abandonment can secure. The philosopher is so forced to recognize principles which long study might not detect. And the naturalist even will stumble upon some new and unexpected flower or animal."[100] *Improve* was one of Thoreau's favorite verbs, but he meant something very different from either the agricultural improvers or their industrial antagonists. "I hate the present modes of living and getting a living," Thoreau had written in 1855, and for him *improvement* meant living rather than getting one.[101] It meant being present, aware of the birdsongs, the taste of the river, the character of the breeze; it meant being alive and sensitive to one's living surroundings.[102] It meant, in July 1859, making a map.

Thoreau began by tracing the outlines of Baldwin and Perham's map and then reusing them for his own.[103] But the two maps' similarities end at the sketched-in river's banks, for Thoreau's is alive with a riot of thousands of tiny notations—most in variously hued pen's ink, but some, ghostly, in pencil—including particular comments about the river's current: "shallow and quick" in some places, "sandbars and grass," "soft banks," in others. And there's a sense of the river's human use: Thoreau notes, just downstream from the Turnpike bridge, "1st cottage," and just a little below, the "boat pl.," from where he and Channing began their explorations. Elsewhere he marks where the good swimming holes are. He points out the cultural and historical geography of the river: where an old hay bridge once stood, where one might find freshwater clams, the monument to the battles of Lexington and Concord in 1776. And he carefully details, in scores of places, what types of plants grow where: individual oak and ash trees, polygonum, bulrush, and many others.[104] This is all a side of the sharp-eyed Thoreau that gained his townspeople's respect as an exacting surveyor, but

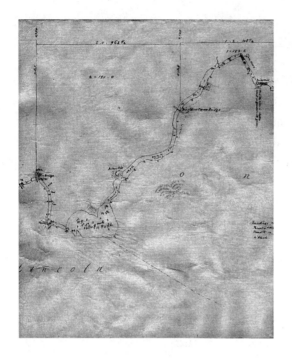

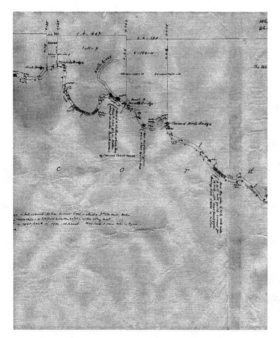

FIGURE 3. Henry David Thoreau, *Plan of Concord River from East Sudbury & Billerica Mills, 22.15 Miles, To be used on a trial in the S.J. Court, Sudbury & East Sudbury Meadow Corporation vs. Middlesex Canal, Taken by agreement of Parties, By L. Baldwin, Civil Engineer. Surveyed Drawn by B.F. Perham. May 1834.* ca. 1859 (detail of rolled survey). Courtesy Concord Free Public Library.

Thoreau was after something wilder: his map seeks to plunge its viewer headfirst into the chaotic mystery of a river called Concord, and it all begins in a cloud of cartographic disorientation with Thoreau's refusal to give the map a coherent name.

The map is overwhelmingly long—seven-and-a-half feet—which means that its viewer actually has to walk its length, must reenact, in miniature, an afternoon's boating trip. It's a demanding map, and to get a grip on it, one must focus on the river itself. It's the only narrative thread immediately apparent, and it's not until the river's conclusion, when it crashes into Talbot's dam, that we stumble upon a title cordoned off in the lower right-hand corner: *Plan of Concord River from East Sudbury to Billerica Mills, 22.15 Miles, To be used on a trial in the S.J. Court, Sudbury and East Sudbury Meadow Corporation vs. Middlesex Canal, Taken by agreement of Parties by L. Baldwin, Civil Engineer. Surveyed & Drawn by B.F. Perham, May 1834*. It's the exact title of Baldwin and Perham's map—except in theirs, the title is front and center, printed in can't-be-missed semaphoric letters, the first thing to seize a viewer's attention. Thoreau then gives to his map a second easily overlooked, even smaller title, this one in a box at the map's center: *Plan and Profile of that part of Concord River between Sudbury Causeway & the Canal Mill Dam in Billerica, Surveyed & the level taken in October 1811, and to be used in The Supreme Judicial Court, in an action then pending between D. Baldwin & J.L. Sullivan, pursuant to agreement of the parties. By L. Baldwin*. Apparently, Baldwin had *also* mapped the river in 1811, for an earlier dispute over the river's height—a fact of which the Baldwin-Perham 1834 version makes no mention.

Directly above this earlier title, a header made up of twin sets of statistics from 1811 and 1834 runs across the map's entire top. The data the figures represent—the river's fall over the course of its length—are easily ignored, the sorts of scientific facts that "may dust the mind by their dryness," as Thoreau complained, but they're also oddly significant because they aren't even close to agreement: the river fell 4 feet 3.9 inches in 1811, but only 2 feet, 10.38 inches twenty-three years later— about 60 percent off the earlier mark.[105] They're measurements divided against each other, and both are cocksure of their accuracy, though both can't be right. As it turns out, Baldwin had been hoodwinked back in 1811 when the Billerica dam's owners, none too pleased at the thought of a survey, cheated. Before Baldwin could set up his instruments, they threw open the floodgates, thus artificially inflating the river's fall and, they hoped, clearing the way for capitalism's rising tide.

Baldwin's folly was public knowledge by the 1850s, but Thoreau refused to let the point go.[106] Instead, he drove home the tenuous connection between map and terrain by including both the 1811 and 1834 statistics, thereby creating a three-in-one palimpsest of contradictory cartographic representations. He wasn't just concerned with the inaccuracy of one map but also in mapping how easily reality can be distorted and its misshapen likeness taken for the truth. To guard against the chance that anyone might misread his fantastic map of conjoined titles and twinned measurements, Thoreau came up with a raving-mad scale to rule his landscape: "60 rods to the inch & B[aldwin]'s is 2/3 of a division of my scale longer than Perham's."[107]

It might be tempting to see Thoreau's river survey as a satirical antimap, a snide rejection of disciplinary pretension that leaves its viewer wallowing in relativism—how could anyone measure anything with a tripolar scale whose differing notions of what an inch is babble over each other? But irony is only one of the tropes that Thoreau drew on. He was always more interested in improvement than deconstruction, and his countermodern map is affirming in its self-aware subjectivity, its desire to picture Concord as situated in a landscape teeming with life and human usage. This was a political choice. If Baldwin and Perham's Concord River is anonymous and untouched and dead; if, in its stillness, their river is an ahistorical space denying change; if the Baldwin-Perham projection captures a river that can be controlled, then all those notes pinpointing what plants grew where, all those piles of figures and ghosts of surveys past, make of Thoreau's a deep map—a view of an impressively interconnected world where nature, commerce, culture, history, and imagination all grow together—something nonfungible and specific: a full, a wild land living at once beyond and beneath the confined landscape of the town's grasping improvers, both agricultural and industrial, who, despite their superficial differences, ultimately agreed that the best use of a river is to turn a profit.[108] Its resistance to the utilitarian clarity of Baldwin and Perham is the same resistance with which all living things meet the violence of abstraction: it won't sit still but demands that you walk it, puzzle over it, bend close, look from afar, and leave it, a bit bewildered. How else could you map a river—long a metaphor for change itself—that, as Thoreau put it in *A Week on the Concord and Merrimack Rivers*, "steals into scenery creating and adorning it, and is as free to come and go as the zephyr"?[109]

Perhaps wildness is always a trespasser disrespecting the artificial boundaries of power. "They who laid out the town should have made the river available as a common possession forever," Thoreau wrote near the end of his life in his great anticapitalist essay "Huckleberries"; and he used all his surveying skill to stake a claim for the bulrush and the ash tree, for the bathers and those who hunted for freshwater clams, for himself and for all Concordians in his map.[110] The Concord persuaded him to use his cartographic training as a means to protest the privatizing of the public goods that ought to benefit every living thing, to disavow the cheap utopian assurances of individual gain so dearly bought. Call it liberation cartography: "I find that I have a civil right in the River," Thoreau had written in 1853, and in 1859 he plated those claims on a map.[111]

8

> Mr. Thoreau
> If you are not engaged to-day I would like to make an excursion with you on the river. If you are, some other day next week.
> W.E.C.[112]

It had been a long summer, a *good* summer, and Thoreau was in the best of spirits when he and Channing, perhaps the closest companion he had after the wrenching loss of his brother John in 1842, rowed back to the boat place. Thoreau had been busily engaged, but he knew that any river trip was really just an excuse to spend long hours with a friend. He thought about this as he and Channing dragged his craft from the current when the willow at water's edge caught his eye, and, realizing that something was missing from his map, he bade Channing goodbye, rushed home, and unrolled his seven-foot masterpiece. Right in the very middle—in the spot usually reserved for titles and beating hearts—he wrote: "Soundings, in feet, so much below summer level—which is . . . 2ft 8 inch below the notch in willow at my boat."[113]

9

Boundaries, when encountered perpendicularly to one's direction of travel, are a rude end to one's walk. But if approached indirectly, from the same direction as the line itself, the line suddenly becomes a route, not the end of a story, but a thread through it, and walking turns into a method for linking spaces, thoughts, and subjectivities into a complex, meaningful essay. Looked at one way, the witness tree is absolutely an end, a property marker whose nonhuman longevity makes it an authoritative arbiter of truth. But looked at with the side of the eye, it invites one to wander, to blend, to see similarity in difference. "Every tree sends its fibres forth in search of the Wild," Thoreau wrote in "Walking," an observation that would have fit the Concord as well: "It is remarkable how the river, even from its very source to its mouth, runs with great bends or zigzags regularly recurring and including many smaller ones, first northerly, then northeasterly, growing more and more simple and direct as it descends, like a tree," wrote Thoreau in the midst of his river survey.[114]

Like trees, rivers can be boundaries. But they can also be allies, and part of what attracted Thoreau to his lifelong companion was its untamable wildness and blithe willingness to help him live in his dream landscape made real: too often, during his terrestrial rambles, Thoreau would have to interrupt his walking and his thinking when abruptly he came to a fence, and he fumed about them, dreaming of "a people who would begin by burning the fences and let the forest stand."[115] But every spring, when it rushed along at its flood-drunk liveliest, the river snatched scores of white pickets and weathered rails, and carried them downstream on its current until Thoreau, afloat in his rowboat, plucked them out of the deluge, took them home, and split them for kindling.

Concord means "harmony."

Nevertheless, boundaries are necessary. They give us meaning. Besides, there's a danger in arguing that no boundaries should exist, the same danger that commoditization and abstraction presents: homogeneity. Place is powerful only because it isn't exchangeable for any other blank space on the map. But to fetishize place, as so often happens in the wider world of environmentalism, can also be a trap. *Place* can all too easily call up images of blood and fatherland, the sort of nationalist and xenophobic environmental rhetoric that has long been so effectively used—*is* no doubt being so effectively used as you read these words—to purge landscapes of people arbitrarily deemed unfit. All places are not good places. All place making is not equally valuable.

We have little choice but to draw boundaries, and this, I think, is where Henry David Thoreau, surveyor, still has something to say to us: "*Who* are we? *where* are we," an overwhelmed Thoreau wrote from the wind-and-cloud-polished peak of Maine's Mt. Katahdin in 1846, the year the Mexican-American War erupted, the year he went to jail, a year when he lived in his cabin at Walden Pond.[116] He never found an answer to those two questions, nor did he ever really want to. He was a good surveyor, and he saw that who we are depends on where we are, that where we are depends on which relationships we choose to honor, that the best places are those always in the process of dawning, that the best boundaries are not the ones that securely wall off one thing from another—art from science, history from literature, politics from land, people from economies, town from woods, country from city, rich from poor, nature from culture, black from white—but the ones that invite transgression so that we all may awake to a new day in a wild, enchanted world. This is the view from Concord.

"So many autumn, ay, and winter days, spent outside the town, trying to hear what was in the wind, to hear and carry it express," he wrote in *Walden*.[117] This is what Thoreau heard when he listened carefully to the breeze sliding over the willow's strings as he prepared for a surveying trip down the Concord in those baked-summer days of 1859:

snatches of a tune poised in haunting counterpoint to the dissonant rumbling of the carts on their way to market and the low grinding of mill wheels on the river's banks, clearer than the muddy prayers of the agricultural improvers trying desperately to wring increased profits from their land—descending phrases of capitalism's ultimate poverty, all: there is no reason any of us must live in a world of violence, commoditized into blandness; no reason any of us must forfeit our creativity, our peculiar genius, our health, our lives, to get a living. No reason we must steal from one another. Life, wildness—all in sympathy, all struggling together for mutual sustenance against hierarchy; even the willow is involved in this ecological and political crusade:

> Ah willow, willow, would that I always possessed thy good spirits; would that I were as tenacious of life, as *withy*, as quick to get over my hurts. I do not know what they mean who call the willow the emblem of despairing love. . . . It is rather the emblem of triumphant love and sympathy with all Nature. It may droop, it is so lithe, but it never weeps. The willow of Babylon blooms not less hopefully here. . . . It droops not to commemorate David's tears, but rather to remind us how on the Euphrates once it snatched the crown from Alexander's head.[118]

ACT TWO THE GEOGRAPHY
OF GRACE

Home in the Great Northern Wilderness

> Away from the sounds of roads and the glare of carbon-arc streetlights, it is quiet here. Some would say it is peaceful, but that is not the right word. This land throbs with life in every season and at every hour. And the quiet itself is not truly quiet. In the absence of the noise of jets and air conditioners, internal combustion engines and recorded music that blanket our perceptions in most of the human environments of America, ten thousand subtler voices may be heard.
>
> JOSEPH BRUCHAC, "At the End of Ridge Road: From a Nature Journal"[1]

1

Words, like boundary lines, make meaning of the world, and some words, like some boundaries, bring definition to the brambly landscape of our minds. One such word is *Adirondack*, a place Thoreau never visited — though he could have. He could have made his way in August 1858, along with Emerson, northwest to New York State's Adirondack Mountains, one among a company that included the world-famous natural scientist Louis Agassiz, the artist W. J. Stillman, the poet James Russell Lowell, and a handful of other intellectuals and Adirondack guides — "Ten men, ten guides, our company all told," as Emerson put it down in his poetic reflection "The Adirondacs" (1858).[2] Thoreau could have joined the adventurers in their trip to one of the premier wilderness areas in the United States, then as now, but he didn't, and he was in a huff about it: "Emerson says that he and Agassiz and Company broke some dozens of ale-bottles, one after another, with their bullets, in the Adirondack country, using them for marks. It sounds rather Cockneyish," he sniffed, as if he had been glad to miss such déclassé gregariousness.[3]

He would have loved the trickster wild, though, with its mythical herds of unicorns (when New York was the New Netherlands, the Dutch refused to set foot in the northern mountains for fear of the horned beasts), its history of castles in the air (like Castorland, the late eighteenth-century dream of a French nobleman grown weary of his country's raging plebeian revolution and its utopian *liberté, égalité, fraternité*, who planned to plant two cities' worth of France's endangered aristocracy in the Adirondack wilderness; or the eight townships, named Industry, Enterprise, Perseverance, Unanimity, Frugality, Sobriety, Economy, and Regularity, laid out at nearly the same time as the French demesne along a virtuous Jeffersonian grid, by a slave trader named John Brown, whose name still marks the landscape as well as an Ivy League university in Rhode Island); Thoreau would have loved the Adirondacks' fantastic, nearly fictional past.[4] Instead, he went for a walk through Concord's meadows.

But not Emerson, who had traveled to Follensby Pond, among the most out-of-the-way spots in New York State, to try on, along with his comrades, new lives, to recreate, to chase into a corner the mysteries of human progress. They imagined themselves going all the way back to the very beginning, where American time began when the mythical, first Promethean white man steps into the wild:

>   We climb the bank,
> And in the twilight of the forest noon
> Wield the first axe these echoes ever heard.
> ....
> Then struck a light, and kindled the campfire.[5]

Of course, the Adirondacks had heard human sounds long before Emerson ever dreamed of lighting history's first campfire: early European visitors reported trails worn a foot deep by Native feet, and recent archaeological and anthropological evidence suggests that, since at least 9,000 BCE, the region has been a "location of exchange" for Paleo-Indians, Abenaki, Mahican, Mohegan, Iroquois, Mohawk, Oneida, and Onondaga.[6]

Yet, in a way, Emerson's fantasy wasn't all that wrong: the Adirondacks then, as now, have a shadowy history, and it has never been entirely clear what, or even where, they are.[7] It's hard to even know where to begin an essay, like this one, that seeks to tell their history, for although they do have a christening date—February 20, 1838, the day that geologist Ebenezer Emmons defined a mountain range somewhere "in the neighborhood of the Upper Hudson and Ausable River" as *Adirondack*—Emmons himself wasn't entirely sure what he was naming, or what the name even meant.[8] When he came to the mountains, he saw range upon named range, all of them different: the Black or Tongue, the Kayadarosseras, the West Moriah (named for the place where Abraham was to take the life of his son), and the Clinton. This last group, named for the Erie Canal–building governor of New York, might better be called Adirondack, wrote Emmons, though he never settled on one particular toponym—in part because *Adirondack* might not mean anything at all.[9]

Emmons had wanted the region to preserve the memory of a "well known tribe of Indians who once hunted here." "It appears," Emmons continued, "that the Adirondacks or Algonquins in early times held all the country North of the Mohawk, West of Champlain, South of Lower Canada, and East of the Saint Lawrence as their beaver hunting grounds, but were finally expelled by superior force of the Agoneseah, or Five Nations. Whether this is literally true or not, it is well known that the Adirondacks resided in and occupied a northern section of the State."[10] He liked the romance preserved by *Adirondack*, but his history is lost, and the uncertainty of the last sentence, "whether this is literally true or not," undermines the confidence of the first. Indeed, *Adirondack*, as far as has been determined, was never an Algonquin word, but a Mohawk one—*atirú:taks*—an insult meaning "the eaters of trees," a slur making fun of one's poverty (as in "you're so poor, you eat tree bark") that the Mohawk likely hurled at their fur-trade adversaries, which may have included various bands of lower Ottawa River Algonquin, Montagnais, and maybe even French fur traders—but no one really knows.[11] It seems that no American Indians ever referred to themselves as Adirondacks, although they certainly gave the Adirondacks names of their own: Wawobenik in Abenaki, and in Mohawk Tso-non-tes-kow-wa ("the mountains") or Tsiiononteskowa ("the big mountain").[12] There were doubtless many others. It was the Dutch who, spooked by rogue unicorns, caught wind of a mysterious people called "Adirondack," and it was they who put the name down in the written historical record where Emmons one day discovered it and carried it in his head as he trekked through New York's mountain wilderness before planting it atop the state's highest peaks.[13]

Nevertheless, as the nineteenth century aged, the banner *Adirondack*, once flying over the Clinton Range only, unfurled and settled upon its neighbors, before slowly drifting downhill to the surrounding woods and lakes, and so eventually came to signify an amorphous area known throughout the century as the Great Northern Wilderness, or penned on maps as the mysterious Wild Unsettled Country.[14]

Adirondack, wilderness, wild, unsettled: the four words were indefinite and interchangeable and indistinguishable in the nineteenth-century New York woods—all of which suited Emerson just fine; he had come to the Great Northern Wilderness looking for the blurred edges of sympathy, not the hard outlines of distinction. "The greatest delight which the fields and woods minister," he had written more than two decades before, "is the suggestion of an occult relation" between the human and sylvan world.[15] So when Emerson, whose body awoke to the downy, perfumed joy of lying on a mattress made up of evergreen boughs and the sharp focus of eyes chiseled open by the dawn's cold, when Emerson discovered an enormous "patron pine . . . fifteen feet in girth" near his camp, it became a witness to the transformation that would overcome the transcendentalist, a shedding of cares, obligations, and social commitments that felt, for all the world, like returning to childhood: "in the woods is perpetual youth," he had written as a younger man.[16] Shaded by his pine, Emerson felt the line delimiting his human body etherealize into the self-aware wild: "The clouds are rich and dark, the air serene, / So like the soul of me, what if 't were me."[17]

And yet, just here, at this moment of ecstatic union, progress interrupted, and the poem takes a turn:

> Two of our mates returning with swift oars.
> One held a printed journal waving high
> Caught from a late-arriving traveller,
> Big with great news, and shouted the report
> For which the world had waited, now firm fact,
> Of the wire-cable laid beneath the sea,
> And landed on our coast, and pulsating
> With ductile fire.[18]

On the shores of a forest-fringed wilderness lake, the flashing news of the transatlantic telegraph, the first time the New World spoke with the Old, jolted Emerson from his communion with nature and into the swaggering pride of accomplishment. "The lightning has run masterless too long; / He must to school, and learn his verb and noun," sung the poet. "We praise the guide, we praise the forest life; / But will we sacrifice our dear-bought lore / Of books and arts and trained experiment, / Or count the Sioux a match for Agassiz?" The answer came back from beneath the boughs of the forest itself: "witness the mute all-hail / The joyous traveler gives, when on the verge / Of craggy Indian wilderness he hears / From a log-cabin stream Beethoven's notes / On the piano, played with master's hand."[19]

Emerson had come to the Great Northern Wilderness hoping for discovery, and perhaps "The Adirondacs" is ultimately a triumphal history of humanity's growing control of elemental nature, from the campfire lighting the wild to the telegraph's pulsating ductile fire. Yet it was Emerson who had written so influentially, "In the woods, we return to reason and faith. . . . In the wilderness, I find something more dear and connate than in streets and villages."[20]

Words are like boundary lines, and they help give meaning to the world in part by fixing a particular history in place, often at the expense of other, alternate histories. Today, many often think of the Adirondacks as simply one of America's first wildernesses—"Forever Wild" has been its tagline since 1894—but what strikes me hardest about "The Adirondacs" is how unwildernessy the landscape it describes is, how full of people, noise, newsworthy events, music, how full of human culture Emerson filled the wild woods ringing Follensby Pond. Cocking an ear to his patron pine, Emerson knew that humans have long looked to trees for guidance, as boundaries, as arbiters of good and evil. Yet he refused to resolve "The Adirondacs," refused to school his reader in whatever it was he learned in the woods, and although it may indeed be a tale of celebration, the poem ends with the teasing suggestion that much of the story has gone untold. "Nature, the inscrutable and mute," remained a Sphinx, a witness with a history it refused to tell.[21]

Words and boundaries help us to know our world, but they can also hem it in, as Thoreau found on the Concord; each can laze into easy, unreflective definition, denying us the rich patrimony of our past. This is an essay, an inquiry twenty years in the making, into the history of two words, *Adirondack* and its ten-letter twin *wilderness*, and the nineteenth-century landscape they describe. These are words, we should not forget, that contain multitudes, and if we're to avoid winding up orphaned in a rootless present, we had better, like Emerson, consider the trees.

2

A good opening question: what is wilderness?

The unobjectionable answer is that wilderness is empty, the way nature was before humans. Wilderness is natural. Both the law—which, with the 1964 Wilderness Act, defined wilderness as "an area where the earth and its community of life are untrammeled by man, where man himself is a visitor who does not remain"—and the ivory tower have long agreed on this point, usually in celebration.[22] But in the 1990s, a few cutting-edge revisionist scholars and cultural critics turned to the idea of wilderness with the keen razors of critical irony, and, after cutting deeply into the intellectual history of the word, they discovered something startling. Supposedly empty wildernesses, like the Adirondacks, have histories, *human* histories, and are "quite profoundly" human creations, in the words of the most eloquently critical and influential historian of wilderness, William Cronon.[23] It's hard, Cronon pointed out, to find a patch of earth that has ever been untouched by humans, and the pristine American emptiness that the Wilderness Act enshrines was in almost all cases produced by Indian removal, foreclosure, anti-squatterism, policing, fire suppression, and restoration ecology.[24] It seems that there are few spaces as intensively manufactured as untouched, natural wilderness.

The tragedy of wilderness is that all that howling emptiness that many of us now celebrate obscures all that violent history of emptying. And because, as Cronon argues, wilderness is the normative intellectual foundation of American environmentalism (wilderness is the way nature *should* be) its "pervasive" and "insidious" nature has inflected how we see the entire world: wilderness is pure and good, while things human—our farms, and towns, and cities; the places we all live—infect whatever they touch and can be redeemed only by a paradoxical return to an antihuman emptiness.[25]

Since it was the Great Northern Wilderness (forever wild!) that taught a great many nineteenth-century Americans about the wild—it was, after all, the wilderness area closest and most accessible to the popular-taste-making major cities of the Northeast—perhaps the Adirondack trees tell of what follows eviction: abrupt historical silence.[26]

3

"I do not often speak to public questions," Emerson had boomed from up on his lecturer's podium in 1854, four years before he tucked himself into the woods of the Great Northern Wilderness. "They are odious and hurtful." He was a poet and a scholar and preferred to be left in the

FIGURE 4. Thomas Cole, *Home in the Woods*, 1847. Courtesy of Reynolda House Museum of American Art, Affiliated with Wake Forest University.

quiet of his study alone with his thoughts, but the decade of the 1850s, rung in when the great Whig compromiser Henry Clay steered through Congress the Compromise of 1850—a bundle of acts that included a strengthened Fugitive Slave Law, a law that could be interpreted to virtually outlaw abolitionism—this was a decade when even retiring scholars had to take a stand. "I have lived all my life without suffering any known inconvenience from American Slavery," Emerson continued, "until the other day," when a Georgian slave named Thomas Sims, who had stowed away on a North-bound ship, was arrested in Boston, sent back to the South by a Massachusetts judge, and finally beaten bloody by his master upon arriving in Savannah.[27] It was more than Emerson could take.

When the famously militant abolitionist John Brown, who shared nothing with the slave trader and Adirondack dreamer but a name, came to Concord in 1857, fresh from the 1856 killing spree in which he supervised the murder of five pro-slavery Kansans—and when he came again in 1859, months before his attempt on the federal arsenal at Harper's Ferry—he and Emerson spent long hours together, deep in conversation. Then, when Brown was caught, tried, and hung after his famously botched arsenal raid, Emerson was one of the few who rallied to his defense. Brown "will make the gallows as glorious as the cross," said the Sage of Concord.[28]

So when Emerson set foot in the Adirondacks in 1858, I have to wonder: was he hoping to run in to Brown? After all, Brown had moved, in 1849, to a house, not far from Follensby Pond, in the town of North Elba. He was drawn, in part, by the mountains' remoteness (Brown owed a great deal of money to a great many people), as well as by the beauty of their evergreen woods. But what pulled most at Brown was the radical gravity of a woodland African American settlement whose members were cultivating, with hoe and ax, a social revolution meant to heal the twisted American body politic of its slavery, not through violent, heroic excision, but with something far more holistic and humble—with care.[29] Maybe Emerson was hoping to meet the extraordinary settlers themselves, pioneers who had left cities and towns throughout New York to make a home in Essex and Franklin Counties, in the heart of what today's wilderness adventurers know as the Adirondack High Peaks region.[30] They had settled communally in a few main nodes, beginning in the mid-1840s. The biggest encampment was around North Elba's outskirts and may have been known as Timbuctoo, an echo of the Mali Empire's ancient capital, a powerful center of civilization when life in much of savage Europe was nasty, brutish, and short.[31] A little farther north was Blacksville. Freeman's Home, yet another settlement, was somewhere off to the west.[32]

In the beginning, before the word, was a thought. In this alternate history of the words *Adirondack* and *wilderness*, there was just such a seed, and it was broadcast to the world on August 1, 1846, from the small, central New York town of Peterboro, nearly 150 miles west of North Elba, by the wealthy landowner, influential abolitionist, and all-around dissenter Gerrit Smith, who would evenly divide, into 40-acre parcels, 120,000 acres of his land, the majority of it in the Adirondacks, for redistribution, at the token price of one dollar per lot, to three thousand black families.[33]

Smith was a complicated man. He was the resentful scion of Peter

Smith—a former partner of fur titan John Jacob Astor—one of the United States' first millionaires, who, every summer, traveled into New York's interior to trade with bands of Mohawk, Oneida, Cayuga, and Seneca Indians for their furry brown bundles of top-hat-destined beaver skins. The elder Smith bled an enormous fortune from the Adirondacks, and Gerrit was delivered into a world that promised him the towering privilege that lies in every bulging wallet. He was also the sole inheritor of the Smith fortune—about $12 million today, in addition to almost a million acres of New York real estate, a great deal of which was in Wild Unsettled Country. Every time he paid a bill, then, Smith's fingertips brushed against the Great Northern Wilderness. But Smith's soul was never in business: he wanted to write, and think, and reform the corrupt world that crowded his rural doorstep.[34] He threw in his lot with the full range of nineteenth-century reform movements, from temperance to women's suffrage, and was what was known as a come-outer: a person who forswore all established organizations—especially those that supported slavery.[35] If there was one thing that Gerrit Smith loathed more than his father's money, it was slavery.

Although New York had a legacy of relatively progressive civil rights for its time—New Yorkers had emancipated their adult slaves in 1827—the state's white citizens squelched efforts to enfranchise all African American men, and in 1846 decided that a black man needed to prove that he was worth at least $250 to vote, a hurdle too high for many. This was an intolerable turn of events for Smith.

Thought, if it's hot enough, can set a simmering soul to boil, and Smith's mind was stoked by the words of William Lloyd Garrison, one of the most influential and radical white abolitionists of his time. Garrison is famous for his strident abolitionist newspaper, the *Liberator*, and in it he developed a militantly nonviolent call for "universal emancipation" from all worldly hierarchies—white over black, man over woman, rich over poor, priest over congregant, ruler over ruled. All power, except for God's, is always wrong, in Garrison's anarchistic view, and so he also advocated for a complete rejection of all government, all mainstream politics, all attempts at working within the system.[36] Smith found the wild logic of Garrison's thinking thrilling—

though he shied from following Garrison down the path to anarchy, arguing instead that abolitionists ought to take the government from within.[37] For Smith was ultimately a man of faith in bedrock institutions, like the Declaration of Independence and its self-evident truth that all are created equal. Didn't that mean, asked Smith, "that no man is to be excluded from the rights of manhood"?[38] And couldn't the gift of productive land, even if it was initially worth less than $250, eventually open the poll's doors?

There were dozens of ways that Smith could have joined the abolitionist cause; indeed, he tried many of them, from founding antislavery political parties to direct action. But for Smith, land redistribution must have been an obvious choice, because rural New York was a tinderbox of radical, anticapitalist agrarianism. In the early 1840s, a wildfire of tenant-farmer activism aimed at abolishing private property and redistributing land from the haves to the have-littles, set New York State ablaze, which resulted in a new political party, the Anti-Rent Equal Rights Party, armed unrest, and the declaration of a state of insurrection by New York's governor when, in 1845, Delaware County's undersheriff was killed by Anti-Renters.[39]

Those flames danced their way through Smith's Madison County and, one day in 1844, even blackened his own front door when an incendiary open letter, published in the newspaper *People's Rights*, landed in his lap. The letter was penned by the all-around working-class radical George Henry Evans, a labor leader, antislavery Free Soiler (he coined the term), and editor of class-conscious newspapers. Evans believed that everyone was entitled to a small farm, and he understood that vast individual wealth always grows from the misfortune of thousands. Large landowners, he argued, not only ate up more than their fair share but also, in blocking access to the land, forced would-be farmers into the wage slavery of industrial labor. So it wasn't much of a leap for Evans to publicly paint Smith as nothing less than "one of the biggest Slaveholders in the United States."[40] It was an acidic line that Evans wrote, intended to cause public pain, and Smith, blistered, knew it. But he also knew Evans was right, and he replied to Evans's open letter with the germ of what would grow into his Adirondack plan: "the individual owners of large tracts of farming land," elites like himself, "should divide them into lots of say, forty or fifty acres, and then give away the lots to such of their poor brethren as wish to reside on them."[41]

The radical potential of agrarianism has always been its promise of independence—one can never be free if one must rely solely on another for life's basic necessities—and it helps explain Thomas Jefferson's zealous faith in the democratic potential of those who labored in the earth. And yet, in the sixty-five years since Jefferson had outlined his thoughts on political ecology, much had changed, and a competing, aristocratic, largely Southern agrarianism theorizing that leisure—the leisure to think and write and cultivate the finer aspects of Western culture—not labor, best ensured the American experiment.[42] By the 1840s, this nondemocratic, *Gone with the Wind* brand of agrarianism dominated Southern political theory, and in Jefferson's South, aristocracy depended utterly upon slavery. "Every plantation," argued the South Carolinian senator, former vice president, and foremost theorist of slavery John C. Calhoun, "is a little community. . . . These small communities aggregated make the State in all, whose action, labor, and capital is equally represented and perfectly harmonized."[43]

Meanwhile, Northern agrarians—including Gerrit Smith—were busy tying Dixie's slaveholders to both environmental and moral degradation, and they began to argue that unhealthy land, unhealthy bodies, and unhealthy societies were all inextricably linked.[44] In 1844, the abolitionist newspaper *Liberty Tree* ran an article, addressed to Northern abolitionists, by Kentucky senator Cassius Clay. In it Clay argued that human bondage "impoverishes the Soil and defaces the loveliest features of Nature." In a modern age, he continued, distinguished by steam power, he and his fellow Southerners were "living in centuries that are gone. . . . In the South where cotton and tobacco once rewarded the husbandmen, can now be seen sterile pine groves, clay banks and naked rocks."[45] Put simply, many in the North started to intuit that landscape and labor system couldn't be unbound from each other, that slavery and free labor were environmental regimes, and that the profits of slavery relied on stealing the earth's fertility quickly and thoroughly while throttling up the demand on black bodies to produce ever more, before wasted lives and spent soil buried the bottom line.[46]

Nowhere was the twinned critique of environmental degradation and human exploitation carried further than in the powerful current of socialist communitarianism that ran counter to mainstream politics, a current that would find its way into the Great Northern Wilderness. By the time the pioneers started arriving in the Adirondacks, communes were everywhere: the first secular socialist society had been planted by Robert Owen, the British avatar of socialism, at New Harmony, Indiana, in 1824, and soon thereafter nineteen other Owenite communities sprang up around the country.[47] Massachusetts, home to both transcendentalists and radical abolitionists, became a hotbed of socialist fervor, giving rise to Brook Farm, a transcendentalist and Fourierist community; Fruitlands, an anarchist Christian association; the Northampton Association of Education and Industry, an abolitionist commune run along Garrisonian lines; and the peace-loving, anarchistic Hopedale Community. All of these were founded in the early 1840s, and all were either biracial abolitionist societies or organizations that strongly supported full racial equality—and that's to say nothing of the black utopias that appeared in the North and Canada.[48]

Though the abolitionists and anarchists and radical reformers and communards were never part of one single coherent philosophy, there was a remarkable degree of intellectual overlap centered on the mutually sustaining relationship between human society and nature. Every association—from the rural Brook Farm and Fruitlands to the more urban New Harmony and Northampton—understood that cultivating a just society required rethinking the mode of relating more generally, beginning with the soil. It meant that one could not accept domination on the one hand and advocate for equality on the other: people and the land, fused by labor, were two faces of a dialectic, and it was impossible for one to be free while the other was chained. A degraded landscape meant that those living on it must suffer. Likewise, a people driven to work with the lash found themselves tending fields that bled in sympathy. Nor was space a blank canvas, or, even worse, a bank of natural resources to be drawn down, but a living thing that formed and reformed even as it was being made and remade.[49] Call it utopian agrarianism.

Or rather, not utopian agrarianism—a futile effort of cultivating nowhere—but eutopian agrarianism: the practice of making a place good.

Together, these eutopians theorized an ecology of freedom characterized by a desire to associate; to grow a beautiful society through fulfilling farmwork that would also enhance the fecundity of the earth; to create goods, not commodities, but goods, which made one's life richer through form and function and the pride of a job well done; to join hands and stand on an equal footing; to form a landscape bearing witness to the truth that fruitful lives and fruitful societies and a fruitful earth are all necessary for each other's safekeeping. Rather than the heroic individualism of Jefferson's white yeoman farmer, or the narrow community of white Southern aristocracy, the eutopian agrarians were guided by a subversive ecology of mutual aid.

Back in New York State, with a million acres weighing down his books, sat Gerrit Smith, his intellect roiling with possibility. If he needed any further inducement to redistribute his land, he found it in the communitarian impulse of his day, for he was not only aware of America's communal societies but also intimately connected to them. John Collins, who knew Smith from the American Anti-Slavery Society, bought 350 acres in Skaneateles, right in Smith's backyard, in 1843 to start his own very serious "No-God, No-Government, No-Money, No-Meat, No-Salt and Pepper" commune.[50] And the Smiths were personal friends with John Humphrey Noyes, who in 1848 founded the Oneida Community in Madison County—the same county that Smith lived in—one of the longest running and most successful of the nineteenth-century communes.[51] Perhaps it was with Noyes's encouragement that Smith wrote letters in praise and support of the Hopedale commune.[52] Perhaps not. In any case, a eutopian spirit worked its charms on Smith.[53] "Happy, thrice happy, will it be when land shall be no more bought and sold!—when, like salvation, it shall be free, without money and without price . . . ! Then, too, there will be no slavery.—Before the reduced-to-practice theories of land-reform, slavery would disappear as surely, and as speedily, as the mists of morning before the rising sun. Apportion the soil equally among its equal owners, and there would be no room for slavery."[54]

Abolition, agrarianism, eutopia—this was Smith's hope for the Great Northern Wilderness.

4

Work and wilderness: surely, these two glare at each other across an intellectual clear-cut. In our modern day, one is either an environmentalist, as the historian Richard White memorably put it, or one works for a living.[55] Indeed, nearly every contemporary environmental issue in the United States—from logging to oil drilling to fishing to nuclear power to global climate change—all are presented as a zero-sum accounting sheet with jobs for the common person competing with nature for elite tree hugger; and mainstream environmentalists, who often thrill to the wilderness as a place for hiking, skiing, or fishing, but never as a place for productive work, are as guilty of promoting a jobs-versus-wilderness narrative as those looking to turn the natural world into a quick dollar. There's no way out: either you're an environmentalist, or you work for a living.

But in the 1840s wilderness had not yet calcified into the notion of an empty, pristine, quiet place for leave-no-trace human visitors. Wilderness was still fluid, a place of contradictions: a sanctuary from persecution, as it had been for the Jews after their Exodus from Egypt, as well as a den for highwaymen and bandits; a place for devilish temptation as well as angelic resolve, as it had been for Jesus during his forty days in the wilderness; an exclusive playground for the aristocracy as well as Robin Hood's steal-from-the-rich populist abode; a scene for anarchic midsummer night's dreams, where trees and animals speak and have souls; and a storehouse for the raw materials of empire. Wilderness could be all of these at the same time.

It could even be a home.

5

Hope is cheap and thought a pastime for the idle patrician if neither cultivates action. Smith had daring and wealth and a plan, but he wasn't ever going to start life anew in the woods—the African American settlers had no need of a "great white leader," he knew, and besides, he was always too ecumenical a reformer to invest himself in one venture. Instead, he played the role of prime mover, to great effect (for instance, he decided that the grantees should be between twenty-one and sixty years old, poor, landless, and sober; for a while, he also actively contemplated giving land to African American women).[56] Even George Henry Evans was impressed.[57]

But the real work of planting and sustaining a wilderness eutopia fell on the shoulders of a picked cadre of elite black abolitionists: Henry Highland Garnet (the fiery pastor of the Liberty Street Baptist Church, in Troy, New York), Dr. James McCune Smith (a writer, critic, and European-trained physician), Theodore Wright (the first African American graduate of Princeton), Charles B. Ray (editor of the black abolitionist newspaper the *Colored American*) and Jermain Wesley Loguen (a runaway slave who would become a bishop in the African Methodist Episcopal Zion Church). These agents combed New York State looking for suitable pioneers, and Smith's agents found everywhere the chosen, people like Charles B. D'Artois, a lineal descendent of France's Charles X; Smith eventually filled 102 pages of a ledger book with names and the exact parcels of land each pioneer received.[58]

Once they had found suitable settlers, Smith's African American comrades took the bare narrative bones of his vision, lifted them from off the columns of his ledger, and animated them with their own still-unrealized, future-perfect prophecy of a time when the nation will have laid racism to rest. Though Smith's commitment to agrarianism was startlingly radical, it was nowhere near that of Wright, Ray, and McCune Smith's, whose hope for the wilderness seemed so nearly real as to have solid form: "In a climate, in which labour is a means for the full and free development of the energies of mankind—in the heart of an almost free state—protected by nearly equal laws—with an equal right to common school education—amidst the friction of advancing civilization—and at a time when the light of science falling upon it has made almost any soil productive—the earth, a free gift, beckons us to come and till it."[59]

Borrowing, when done right, can be artistic. If Wright, Ray, and McCune Smith's declaration sounds familiar, it's because its rhythms broadly echo the martial rhetoric of Manifest Destiny. Yet at the same time, the resemblance is faint because they transposed the tune of western expansion to fit their own needs. The westering impulse was defined by the extension of slavery and the triumph of exploitation, but Wright, Ray, and McCune Smith gave the nation's compass a disorienting kick: in place of the familiar celestial sign of a setting sun guiding white Easterners out to colonize the red West, in the Adirondacks, Polaris-following black pioneers sought to civilize a white North as a preliminary step toward conquering an entire continent's racist savagery.

"There is no prejudice under which we suffer," Smith's agents continued, "which may not be removed, no oppression under which we labour, which may not be meliorated, by a prompt and energetic movement in the direction of this glorious opportunity"—that is, in a northern, wilderness-bound direction. "Once in possession of, once *upon our own land* we will be our own masters, free to think, free to act.... Thus placed in an independent condition, we will not only be independent, in ourselves, but will overcome that *prejudice against condition*, which has so long been a mill around our necks."[60] With the millstone gone, the pioneers could stand upright, "tall, stalwart, hard-fisted, they embody a Hope of the Race," James McCune Smith wrote back to Gerrit Smith, sure that cultivating the land was another way to cultivate both the self and society.[61] For farming in the Adirondack Mountains would throw people of all hues together: "There is no life like that of the farmer, for overcoming the mere prejudice against color. The owners of adjacent farms are *neighbors*.... There must be mutual assistance, mutual and equal dependence, mutual sympathy—and labour, the 'common destiny of the American people,' under such circumstances, yields equally to all, and makes all equal."[62]

Borrowing can also be its own act of creation, and underlying all of this rhetoric of blooming wilderness and mutual aid is something that looks like the socialists' eutopian agrarianism, but that, when transplanted to the Adirondacks, emerged as a budding environmental philosophy in which a working wilderness landscape bore not only political freedom, though it did, or even a freedom more metaphysical, though it did that, too; what was most eutopian about McCune Smith's piney landscape vision of "mutual assistance, mutual and equal dependence, mutual sympathy" was that it promised freedom from the problem of race.[63]

What is race? The word is so common today—we're checked by its boxes at all of life's milestones: at birth, by prospective employers, upon death—that we often forget that *race* means more than the color of a person's skin, more than the origins of a person's ancestors. Race is the notion that humans can be sifted into groups that are biologically distinct from one another, and that those biological distinctions trump everything—culture, education, politics, perhaps even evolution. Races are the human equivalent of animal species. Just as different species can be told apart on the basis of physical difference—a glance distinguishes a coyote from a fox, though both may be of the family Canidae—racial thinking assumes that outward appearance is the key to unbridgeable inward difference. To be defined as white, then, means much more than to have skin lacking melanin—but just how much more has long been a fraught problem, for the simple fact is that race has no scientific evidence. There are no separate species of the human family. We're all equally human.[64]

Nevertheless, the idea that race is real continues to persist because racism has willed it into existence, because without the notion of objective races, racism collapses into mere prejudice, into easily ignored, vile opinion.[65] And so, beginning with Thomas Jefferson, some of America's finest scientific minds, in order to justify a racist society, have bent the arc of their thinking toward crafting a world in which black was naturally inferior to white. "Whether the black of the negro resides in the reticular membrane between the skin and scarf-skin," wrote Jefferson, "or in the scarf-skin itself; whether it proceeds from the colour of the blood, the colour of the bile, or from that of some other secretion, the difference is fixed in nature, and is as real as if its seat and cause were better known to us."⁶⁶ With surgical language that can't quite conceal his misshapen, a priori faith in inequality, Jefferson reveals that "black" had nothing to do with skin color and everything to do with what he took "black" to mean: monotonous, ugly, inelegant, odoriferous, requiring less sleep and incapable of suffering (handy traits for slaves), impulsive, lusty, happy, dim-witted, lazy, senseless to art and poetry and refined music.⁶⁷ "Will not a lover of natural history, then, one who views the gradations in all the races of animals with the eye of philosophy, excuse an effort to keep those in the department of man as distinct as nature has formed them?"⁶⁸

By the middle of the nineteenth century, natural scientists were taking up Jefferson's call to segregate humans, and their work would eventually grow into the scientific racism that erupted in the post–Civil War United States. Samuel George Morton's *Crania Americana* (1839), which was based on measuring more than a thousand skulls, lent a sheen of scientific authority to the fantasy of race by seeming to prove that skulls could be grouped into a hierarchy of five different races, with characteristic shapes and capacities—a white supremacist tree of life. Anglo-American skulls, of course, were the most beautiful, in Morton's eyes, as well as the biggest, and so the "Caucasian race" was "distinguished for the facility with which it attains the highest intellectual achievements," the king of the human species. The "Ethiopian race," however, with its small, ugly skull, "is the lowest grade of humanity."⁶⁹ But perhaps Morton's biggest contribution to scientific racism was his suggestion that different races belonged to exclusive geographic regions. And he helped lend credence to the theory of polygenism—polygenists believed that each race was a separate thought of God's—by arguing that each race was a distinct human species particularly suited to separate geographies.⁷⁰ People with dark skin, so the thinking went, belonged naturally in the tropics.

Perhaps scientific racism would have sputtered to an inglorious stop had not Emerson's Adirondack companion, Louis Agassiz, America's celebrity scientist to whom we owe the discovery of ice ages, become a fast convert to Morton's thinking in 1846—the same year Smith announced his plan for ending racism. From Morton, Agassiz inherited his belief in polygenism, as well as the revelation that each race had its own separate place on the earth; by 1850, Agassiz was sermonizing his creed from the pulpit of Harvard's Lawrence Scientific School, giving race and racism the solid intellectual foundation of Ivy League prestige.[71]

Black was biological. Black was innate. Black was inferior. Black was geographically determined. Such was the climate into which the Adirondack pioneers stepped. But if people with dark skin could cultivate the northern (white) woods, if they could thrive amid the snow, then they could show that what America's scientific elite was beginning to claim was natural and biological was, as James McCune Smith put it, "mere prejudice," a political and social condition that required a political and social remedy.[72]

One can drift far away from the day-to-day on the hot air of theory; yet New York's African American communities found all the lofty thinking about landscape and sweat and remolding the nation to be grounding, even despite the Great Northern Wilderness's reputation for harshness. "I wish the land was in a less rigorous clime," Smith wrote to Frederick Douglass, who was surprised to discover a deed for forty Adirondack acres tucked in to Smith's letter, "but it is smooth and arable, and not wanting in fertility."[73] "The first settlers of this town," Smith wrote at another time, "thought the land equal to the Illinois land, and it did produce well at that time. . . . They tell of raising Rye here 9 feet high."[74] The region caused a certain amount of consternation among the grantees, of course—something less than two hundred actually settled their lands—but it also fortified their spirits: in 1847 thanks started to erupt from across the state.[75] Willis Hodges of the *Ram's Horn* trumpeted the Adirondacks in his paper, while the *Albany Patriot* and the *Impartial Citizen* from Syracuse published firsthand accounts and editorials singing support for the Adirondacks. By 1848—that year when the Treaty of Guadalupe Hidalgo ended the Mexican-American War, putting a keen gleam in slaveholders' eyes as they contemplated an empire of slavery; that year also of socialist revolution in Europe—Gerrit Smith could write that "some twenty or thirty are comfortably settled" on their new land and that "the remainder are preparing to follow them in the Spring."[76] The year's spring saw Hodges selling his interests in the *Ram's Horn* for a team and a wagon and a new life in Franklin County. He moved not without trepidation—having grown up on a farm in Virginia, the son of a free farmer, he knew the challenge of the Adirondacks' thin soil. Nevertheless, in May he led a group of four families and five single men to Blacksville, on Loon Lake.[77] Even Douglass threw his full weight behind the plan. Between 1848 and 1850, at least nineteen articles on the Adirondacks appeared in his paper, the *North Star*, and he exhorted his readers to head north in language that grafted military metaphors onto religion, the pioneer's rosy western hope for the future onto the slave's nightmarish present:

> Advantage should be at once taken of this generous and magnificent donation.... The sharp axe of the sable-armed pioneer should be at once uplifted over the soil of Franklin and Essex counties, and the noise of falling trees proclaim the glorious dawn of civilization throughout their borders.... What a man soweth that he shall reap.... Come, brethren, let it not be said, that a people who, under the lash, could level the forests of Virginia, Maryland, and the whole Southern States, that their oppressors might reap the reward, lack the energy and manly ambition to clear lands for themselves.[78]

One reaps what one sows, and this is why when many of the pioneers and their supporters wrote of axes felling trees, of clearing land and planting grain, they never experienced the involuntary shudder of today's environmentalists when *ax* and *wilderness* wind up in the same sentence. And though they spoke openly of clearing, their vision of a proper human role in nature was fundamentally different from what the promoters of Manifest Destiny meant when they fantasized of forests falling to amber waves of grain.

For by the 1850s, the Adirondack forest was beginning to show the effects of market-based, industrial logging. It wasn't so much that timber barons were stripping their land bare for lumber—that wouldn't start to occur in earnest until after the Civil War—but that large tanneries, which had set up shop throughout the Adirondacks, were systematically targeting the mountains' hemlock, whose bark was the primary source of the tanners' tannin. A large tannery could consume as much as six thousand cords of bark a year (a cord is four feet by four feet by eight feet, or the bed of a pickup truck filled to the top of the cab), and, by 1865, there were close to 197 operations sending their loggers into the woods. The result was that, by the 1890s, the tanneries had cut over more than a million acres of Great Northern Wilderness woodland.[79]

At the same time, the intensely capitalized iron forges and furnaces scarring the eastern cheek of the Adirondacks were consuming charcoal at a voluminous rate. And because charcoal can be made from any species of hardwood, entire forests, seven thousand acres per year during the peak midcentury years of iron production, toppled before the dollar's cutting edge. If tanning left the forest thinned of its hemlock—which occurred in pure, intermittent stands—the forges left entire swaths of the landscape bald.[80]

Finally, the Adirondacks were one of the prime lumber frontiers of the nineteenth century, and they gave birth to innovations that changed the industry, like the log drive. All of this helped to propel New York to the number-one spot in terms of lumber-producing states by 1850, when more than two thousand mills floated vast quantities of white pine, red spruce, and especially hemlock, out of the mountains. By 1870, more than a million logs a year were being cut in New York—many of them from the Adirondack forests.[81]

But Timbuctoo and Freeman's Home and Blacksville were to be places for husbandry, rather than a cut-and-get-out scene of pillage, and in letter after letter, contributors to the *North Star* wrote that settling the Adirondacks would be a sort of homecoming.[82] As one commentator put it, "Forsake the cities and towns and . . . settle upon this land and cultivate it, and thereby build a tower of strength. . . . Forsake the cities and their employments of dependency and emigrate to those parts of the country where land is cheap, and become cultivators of the soil."[83] William Jones, a former slave who had emancipated himself by running away from his master, listened to an address at Henry Highland Garnet's Liberty Street Church, then rose to exhort the gathering: "God bless Mr. Gerrit Smith, and all the Smiths" he began, "come off the steamboats—leave your barber shops—leave the kitchen, where you have to live underground all day and climb up ten pair of stairs at night. To-morrow morning I intend to leave for Essex County to see for myself."[84] Whereas towns and cities were theorized as places of skyscraping vertical dependency—where the poorest lived underground, where, as the grantees from Rochester put it, African Americans dwelled in "the subordinate offices now assigned them in the cities"—in the Adirondacks, everything and everyone lived on a level wide open to the sun, awash in an evergreen breeze, a place where one could breathe free: "Aspire for the soil," cheered the Rochester pioneers.[85]

For the dark-skinned pioneers, the forests of the Great Northern Wilderness were home: "The land is open to them. The land has just as much respect for a black man as it has for a white one.—Let our colored brethren betake themselves to it."[86]

6

All of the literature, both scholarly and popular, revisionist and celebratory, on wilderness has long been clear: wilderness is by definition a white thing.[87] Look at the well-known patron saints of pure American wilderness, and they're universally white (and, down to the person, almost entirely male). Indeed, exposing the whiteness of wilderness was one of the revisionists' most important contributions.

Yet the trouble with the revisionist wilderness critique is that collapsing all notions of wilderness into one single idea articulated by a single, homogeneous group of Americans, leaves no room for anyone else and, perversely, winds up erasing the very alternate histories—including histories of environmental justice—the revisionist scholarship sought to recover.[88] "The most troubling baggage that accompanies the celebration of wilderness," writes William Cronon, "has less to do with remote rain forests and peoples than with the ways we think about ourselves—we American environmentalists."[89] But what about all the Americans who have never, and still don't, identify with Cronon's imperial *we*? Perhaps the rest of us, too, can think about—perhaps we *have* thought about—wilderness; perhaps those left out by Cronon's *we* haven't all always agreed that wilderness should be empty, or is empty. And perhaps not all important wilderness thinking comes from the brotherhood of lily-white wilderness thinkers and their modern-day, mainstream-environmentalist heirs.

Part of the problem is that the favored digging ground of intellectual historians—essays, philosophical tracts, legal documents, newspaper editorials—privileges privilege. There's no black *Walden*, no *My First Summer in the Sierra* penned by a dark-skinned John Muir, but that doesn't mean that the history of environmental thought can be found only in the standard Western tradition of belles lettres, or that African Americans (or, for that matter, anyone else other than the dominant) were ignorant of powerful cultural currents. And if what counts as "intellectual" can be opened to include sources typically ignored by intellectual historians—like the slave song "I'm Going Home"—perhaps we can begin to piece together alternate ideas of wilderness:

> I sought my Lord in de wilderness, in de wilderness, in de wilderness;
> I sought my Lord in de wilderness, For I'm a going home. . . .
> I found free grace in de wilderness, in de wilderness, in de wilderness;
> I found free grace in de wilderness, For I'm a going home.
> My father preaches in de wilderness, in de wilderness, in de wilderness;
> My father preaches in de wilderness, For I'm going home.[90]

Poet and literary scholar Melvin Dixon argued that slave songs like "I'm Going Home," collected from Nashville, Tennessee, in the 1860s, show African Americans working a complicated wilderness—an alternate space far from the killing "pastoral order" of the plantation—into a "geography of grace," of *free* grace, a space to find the Lord and perhaps a long-ago sold-away father, even a home, a place where the Fugitive Slave Laws had no purchase.[91]

What might a more richly researched, more imaginatively inclusive wilderness tradition look like? What if it were more receptive to alternative wilderness ideas and different intellectual trajectories, ones that theorized work as a legitimate way of knowing raw nature (an ecological route historically traveled by African Americans)?[92] What if this more radically inclusive history understood wilderness as a space of liberation that had to be maintained by caring human hands, a place where identity, landscape, community, and memory defined one another?[93] Such would be a creative definition of wilderness that presupposed occupation and laboring, an idea that would have little to do with the notion of the lone, white, male sojourner coming face-to-face with a sublimely empty nature while on vacation. It would be an equation of wilderness with freedom, with community, with home.

7

It took a while for him to get there, but when James McCune Smith finally betook himself to the Adirondack woods, he found salvation.

In late 1846, Smith had asked McCune Smith to pen a few words of an address to African American New Yorkers. It was supposed to be a pitch, lauding the Great Northern Wilderness and exhorting the grantees to claim their lands, but McCune Smith hesitated. He was depressed, feeling beaten, like a lifetime of hard work, work for which he had sacrificed everything, was worth exactly nothing. If anybody in the United States had meticulously groomed himself according to all the fiddly strictures of the self-made man, it was McCune Smith—the Rockefellers and Carnegies, all the Ragged Dicks of later generations, stood shabbily in his shadow. He was the first professionally trained black physician in the United States—but since his color had barred him from studying at any American school, he went abroad and earned a BA, MA, and MD from the University of Glasgow. He was a writer, too—he wrote the introduction to Douglass's *My Bondage and My Freedom* (1855)—and a literary critic (he published a review of *Moby-Dick* in 1856).[94] Finally, McCune Smith was one of the most important and influential antiracist theorists of his day, and he argued that American progress depended on inclusive diversity.[95] But to most white eyes, all of this was colored by the fact of his dark skin. To most white eyes, McCune

Smith was nothing. And so when Smith asked him to help cowrite the broadside, McCune Smith replied: "I have no heart to write it. Each succeeding day, that terrible [intolerant white] majority falls sadder, heavier, more crushingly on my soul. At times I am so weaned from life, that I could lay me down and die, with the prayer, that the memory of this existence should be blotted from my soul."[96] He had stared into the face of hatred long and hard enough for it to nearly break him of hope: "There is in that majority a hate deeper than I had imagined. Caste, the creature of condition, I supposed to be feebler than any strong *necessity*. Yet here came a necessity, the strongest this people knows—a political necessity—and lo! it is weaker than caste! Money is weaker than caste; Political necessity is weaker than caste—to what else will this stiff-necked people yield? Labouring under these views, I cannot write a cheering word & I will not write a discouraging one."[97] McCune Smith was dangerously close to accepting racism as inbred and thus unyielding to every attempt at reason, moral suasion, nonviolence—even self-interest. But still he clung to the bedrock faith that "physical force has no place" in "changing the heart of the whites."[98]

In his despair, he lashed out at Smith and his Adirondack plans; he was trying to cut, and accused Smith of simplemindedly kowtowing to an economic system designed to cultivate inequality: what good was the vote, what good was $250 worth of land if it was ultimately inanimate property, rather than the person, that wound up with political recognition? "My personal influence, manhood, presence at the ballot box is utterly destroyed when the earth-owning oath is thrust at me," McCune Smith raged. "The negro *Man* is merged into the negro Landowner. . . . It is established by our oath, that the vile earth has rights superior to Mankind! That 'the dust of the earth' is the greater, without 'the breath of life.' What horrible mockery!"[99] Wasn't Smith simply playing the game, legitimizing disenfranchisement, hierarchy, and domination with his reformist get-out-the-vote scheme?

In the end, however, McCune Smith choked back the bile, picked up a pen, and wrote the letter. Maybe he sat in his chair, a tall stack of paper within easy reach, and mechanically scratched, revision after revision, synapses firing, muscles twitching out the phrases in a willed act of duty. Maybe he did it out of friendship, or obligation. But I think something else happened: I think he realized that voting was only ever a secondary consideration, and that "'the dust of the earth,'" and "'the breath of life'" by themselves were nothing, but combined they gave birth to something living, thinking, active. That in the union of the two was the birth of history.

When McCune Smith wrote that Smith's plan diminished the human "negro *Man*" into the mere holder of a commodity, the "negro Landowner," he assumed that the Adirondacks were to instill the very sort of individual up-by-the-bootstraps ethic that had proved so disappointing and superficial in his own life. Individual work alone can't kill racism, but as it turns out, the pioneers already knew this. In November 1846, one month before McCune Smith mailed his letter of despair (and three months after Smith had made his proclamation), a committee of willing pioneers from Albany who had heard of Smith's plan contacted Smith with an idea of their own: they wanted him to sell them seventy-five thousand Adirondack acres, for ten cents an acre, which they would then divide into large lots, open to parties of African Americans who would till the soil cooperatively.[100] Though the plan never came to much, it shows that African Americans in the North were already thinking through the logic of Adirondack settlement and leaning toward some sort of communal model. Then, six months later, in the spring of 1847, Smith's fellow agent Charles B. Ray returned from his wilderness tour of inspection with a glowing report of the Adirondack land, finding it "fairer than you [Gerrit Smith] represented it to be."[101] His young daughter was even more enthusiastic, and when she returned from the woods sang out that she was an "Essex Co girl"; the only fault Ray could find was with the plots he and McCune Smith had been given: both were too distant from the hub of activity at Timbuctoo, and they requested a relocation to a parcel with a more "central position"—something nearer the Deacon Iddo Osgood's place, one of the original white settlers of North Elba, a tavern owner, and, it seems, one of the pillars of the rustic society.[102]

What ultimately seems to have changed McCune Smith's mind was the eutopian promise: the individual power of any "negro *Man*," even an exceptional one, would never jam the works of a racist society, but that of a committed community might, and by the spring of 1848, McCune Smith was pluralizing his prose. "There is a good spirit amongst the grantees who have received their deeds," he wrote; "I look with joy to mixing with the strong hardy men."[103] A few months later, a company of pioneers returned to New York City to tell McCune Smith and others of their triumphs and to seek support for a cooperative plan: each settler would pay $1.25 in initiation fees and fifty cents a month thereafter in dues to join a common fund; the proceeds would then be used to survey each member's lot, cut roads, clear land—in short, to ensure community development. McCune Smith liked what he heard so much that he felt "very desirous to go on the good land," and even began fund raising for a good team of horses or oxen, to be held by the pioneers in common for the purpose of shipping and delivering goods, baggage, and people to and from Port Kent—one of the main jumping-off places for the Adirondacks.[104]

But as the days of 1850 grew short and the autumn leaves fell from their parents' limbs in that terrible fall that saw the passage of the Compromise of 1850 and with it the devil-pact Fugitive Slave Act, McCune Smith's beloved first child, his young Amy, died. He was devastated. He wrote Smith with the news: my daughter is dead. But somehow he continued, bidding his hand to move and his thoughts to wend their way from the unspeakable to the possible; just a few weeks earlier, he had been in the mountains, in September, when the sun often shines in a bluebird sky, just before he lost Amy, and he wrote to Smith about meeting sixty pioneers, "of all ages and sex," settled around North Elba, an outpost that, since it harbored runaway slaves, had become officially criminal. Though they were outlaws in the eyes of the federal government, and though money was tight, their spirits were high, and, besides, they were industrious: "They had put up several good log houses. . . . I think more clearing has been done within a year in To[wnship] 12 then in any three years together of late."[105]

What he really wanted to tell Smith in his own indirect, literary way, was that he, a desperate, grieving father, had found a green-tinged grace in the Adirondacks: "I felt myself a 'lad indeed' beneath the lofty spruce and maple and birches, and by the baubling brook, which your deed made mine, and would gladly exchange this bustling anxious life for the repose of that majestic country, could I see the day clear for a livelihood for myself and family." McCune Smith didn't and wouldn't move to his Adirondack land—his ties to New York City were too strong—but standing on his own bit of ground, with the maple and birches and fir as witnesses, he was reborn. He found repose, and he found music: "As we went north thro' township 11 and 10, we found . . . here and there colored settlers making their woods ring with . . . their axe strokes."[106]

John Thomas was one of these settlers. In 1839 Thomas ran from the Maryland plantation of his master, Ezekiel Merrick, after Merrick sold his wife away. Over a period of nine years, Thomas slowly made his way from Maryland to Philadelphia, then to Troy, New York, and finally to Essex County. Illiterate but grateful, in 1872 he hired someone to write a letter to Gerrit Smith, a beautiful bit of correspondence, written in an unhurried, steady hand that adorned the letters with graceful arabesques. Thomas began by praising Smith's "benevolence towards myself, as well as my Colored Brothers generally," before moving on to detail his life in the Adirondacks. Originally given a forty-acre tract, he sold it, "owing to inconveniences of Church and school principles." But rather than leave the mountains, Thomas bought a different plot, closer to his community's center, "which by labor and economy has been enlarged into a handsome farm of two hundred acres; with all necessary stock and farming implements. I generally have a surplus of two or three hundred dollars worth of farm produce to sell, every year." Thomas had made it, and in closing, he testified that owning land, farming it, allowing it to flower had actively changed him: "I have breasted the storm of prejudice and opposition, until I begin to be regarded as an 'American Citizen.'"[107]

That should be the end of this essay. In a good world, the legacy of Adirondack wilderness should be defined as much by its history of social justice as by its pathbreaking role in preservation. But, today, in the early twenty-first century, thousands of hikers and birdwatchers and walkers and snowshoers stroll over land once called Timbuctoo and Blacksville and Freeman's Home with no intimation that it was ever anything but virgin forest. The Adirondacks never became the staging ground for ending racism, and little by little through the late 1850s most of the pioneers trickled back to their New York homes, partly for reasons that James McCune Smith well understood: "Unless . . . I can make enough to secure an income of $400 per annum, I must defer settling in Essex County. . . . Could we get about 200 settlers in North Elba, and then cut off all communication with the city . . . things could be made to prosper," he wrote from his New York City home.[108] There's nothing easy about living in the woods, and it's not as if McCune Smith was alone in his reluctance. Frederick Douglass never moved north, nor did William Wells Brown or Henry Bibb, prominent African American abolitionists, escaped slaves, authors, and owners of Adirondack land, all. Even those among the pioneers who were experts with shovel and hoe weren't prepared for the ecological realities of the Great Northern Wilderness, because the thin soil of Essex and Franklin Counties, combined with the narrow window of seasonable weather (snow can fly in June) meant that Adirondack farming then, as now, was equal parts hard work, good luck, and blind faith; it was even tougher for those who had made their livings as barbers, mechanics, or laborers in New York City. For too many, life in the North Country was too tenuous.[109]

Stony soil and cold weather, however, account for only so much; after all, the pioneer James Henderson reported in 1849 that "there is no better land for grain." "We get from 25 to 50 bushels of oats to the acre," he continued, "and for potatoes and turnips . . . we get from 200 to 400 to the acre.—The farmers here get 46 cents per bushel, cash in hand, for their oats."[110] Lyman Epps lived out his life on Adirondack soil (he moonlighted as an Adirondack guide and cut the first trail to Indian Pass, a trail beloved of today's hikers), as did his son; and various other families stuck well into the nineteenth and even twentieth centuries.[111] Clearly, some pioneers could make a modest living in the wilderness.

But what proved fatal was the arrival, in 1849, of John Brown on his red warhorse. He was penniless and, a failed land speculator and wool dealer, on the run from his creditors. He was also deeply religious, a true believer in biblical violence, a man who believed slavery a mortal sin and who had the dewdrop-pure conviction that he himself was

God's lone, vengeful enforcer.[112] Like Gerrit Smith, he looked forward to God's government, but Brown's was a domineering one that thrived upon force: the chosen were meant to lead, by any means necessary, and there was never any doubt in Brown's own mind that he was chosen. In April 1848, with nowhere to go and mounting debts dogging his steps, Brown headed to Peterboro, met Gerrit Smith, and grandiloquently declared, "I am something of a pioneer; I grew up among the woods and wild Indians of Ohio, and am used to the climate and the way of living that your colony find so trying. I will take one of your farms myself, clear it up and plant it, and show my colored neighbors how such work should be done; will give them work as I have occasion, look after them in all needful ways, and be a kind father to them."[113]

To his own father, Owen, he betrayed another reason for choosing to head north: "There are a number of good colored families on the ground; most of whom I visited. I can think of no place where I think I would sooner go; *all things considered* than to live with these poor despised Africans to try, and encourage them; and to show them a little as far as I am capable how to manage."[114] It is plaintive, that emphasized note, "all things considered": Brown was desperate.

In the spring of 1849, Brown moved his family of nine to North Elba, but not to the 244 acres that Smith had agreed to give Brown on credit—that land was still largely uncleared, and there was no home on it in which to live. So the Browns rented a smaller house from a man named Flanders, and from their new lodgings they paused for a moment to notice "how fragrant the air was, filled with the perfume of the spruce, hemlock, and balsams."[115] This rented house was really the only Adirondack home that Brown ever knew, and the only extended period of time that he ever spent in the Great Northern Wilderness was from 1849 until 1851, when he moved his family back to Ohio. But even during that two-year period, he was traveling throughout the East, defending himself in court against a growing pack of creditors, trying to buy time, and perhaps deploying his charismatic genius for promotion on the pioneers' behalf.[116] In 1855, he once again moved his family back from Ohio to North Elba, into a partially finished farmhouse that had been built for him (it's the John Brown Farm that one can still visit today); but Brown himself didn't stay. Kansas was bleeding in a civil war between pro- and antislavery settlers who were fighting to determine whether the territory would enter the Union as a slave state or a free one, and he felt himself called to the conflict, called to murder five proslavery settlers, which would electrify his reputation as a person willing to kill for freedom. Brown's farm was enlarged in 1857, when a band of Massachusetts philanthropists raised $1,000 for his struggling family, but Brown would see the new land only a handful of times before he left for his raid on Harper's Ferry in the summer of 1859.[117]

And that was the extent of John Brown's Adirondack adventure.

His body, according to his dying wish, was shipped back to North Elba after his execution and buried in ground that he had never spent much time on but cared enough about to spend eternity in. It's clear that Brown deeply loved the Adirondacks—in 1854, while he was in Ohio, he wrote, "My own conviction, after again visiting Essex County . . . is that no place . . . offers so many inducements to me, or any of my family, as that section. . . . I never saw it look half so inviting before."[118] And it's clear that the colony of pioneers meant a great deal to him: he asked constantly about them in his letters, and arranged provisions to be sent their way—he very much wanted the pioneers to succeed as farmers. But it's also clear that he had no use for their pacifist eutopian agrarianism.

What's not clear is how much John Brown meant to the colony. Though many of Brown's biographers, in trying to make sense of why he went to the Adirondacks in the first place, exaggerate the degree to which the pioneers were "his community," and he the community's head, it's significant that Brown is rarely mentioned in any of the letters or newspaper articles written by the pioneers themselves, and significant that he had no luck recruiting pioneers for his Kansas or Virginia raids, though he tried.[119] And McCune Smith, far from seeing the colony as Brown's, feared that the settlers "were a little too dependent upon Mr. Brown's meal bin."[120]

Unwilling to conceive of the Adirondack colony on the terms of its founders, inhabitants, and supporters, unwilling to imagine himself a part of the project of mutual aid and environmentally driven social justice, Brown used the Adirondacks as a redoubt from which to marshal the attention, blessings, and, perhaps most important, intellectual and financial support of the more economically elite radical abolitionists. Spurred on by Brown's magnetic personality and the repellent Fugitive Slave Act, Gerrit Smith found himself enthralled by the immediate possibilities of violence, so much so that he became one of the Secret Six who had foreknowledge—and indeed helped fund—the Harper's Ferry raid; Frederick Douglass, too, knew what Brown was up to, and he found himself bending in Brown's direction. Even the doctor, James McCune Smith, began to feel that blood must be let.[121]

In the end, Brown was right: slavery would be purged only by killing. No matter whether you think Brown a martyr or a madman, he was certainly one of those rare characters, perhaps a world-historical figure in whom is embodied a historical flash point, a point of no return. No garden can flourish without attentive gardeners, but the Adirondack colony's downstate backers had had enough with hoes and axes once Brown showed them a gun. Left stranded, most of the settlers slowly trickled back to their former homes while the wilderness reclaimed an emptied Timbuctoo and Blacksville and Freeman's Home.

It's not fair, I realize, to pin the demise of the Adirondack venture on Brown; but it's worth seeing Brown as the personification of its ultimate failure—and of the failure of radical abolitionism as a whole: the turn to violence at the expense of fundamental social and environmental rebirth.[122] It's all the more unfortunate because after the Compromise of 1850, as both mainstream political parties, Whigs and Democrats alike, committed themselves to the protection, even the extension of slavery, and, what's more, seemed to accept racism as implicitly integral to America; as the political rhetoric grew hotter and the real-life situation of African Americans grew more desperate; as Gerrit Smith and Frederick Douglass and James McCune Smith became convinced that Brown's way was the only way forward and so began to lose interest in the Great Northern Wilderness; as abolitionists increasingly responded to the always-easy seductions of bloodshed; as violence came to be seen as a legitimate political tool and, especially after the war's end, a legitimate economic one (dogs eat one another in the world capitalism built); as Brown's thundering critique of slavery, but not racism, reached its terrible crescendo in the exploding shells on the fields of Antietam, Shiloh, and the Wilderness, and then stammered to a stop when the Thirteenth Amendment put an end to slavery in 1865 and a sharecropper economy swapped golden chains for iron ones; as social justice movements started to fracture along the lines of identity politics (women's rights and antiracism advocates notoriously went for one another's throats in the postwar decades), and as the free market and competitive individualism came to be considered the only real route to freedom, the costs be damned—as all of this came to pass, the nation desperately needed the ecology of freedom that had flowered in the brief Adirondack spring.[123] But by then only a few settlers remained.

8

Definitions have such power because an entire story lives within the chrysalis of a single word, and, if enough attention is paid, one can catch a definition as it emerges, wings still wet, into the world. In 1869, ten years after Brown was buried and only four after a few hundred thousand more Civil War dead were laid beneath the sod, W. H. H. Murray, a never-tiring promoter of the Wild Unsettled Country, was hard at work shoehorning all two hundred pages of his *Adventures in the Wilderness; or, Camp Life in the Adirondacks*, a book that emptied the woods of its human history, into the conjoined words *Adirondack wilderness*.[124] But he paused a moment to consider John Brown in his grave:

> At Keeseville, if you wish, you can turn off to the left toward North Elba, and visit that historic grave in which the martyr of the nineteenth century sleeps, with a boulder of native granite for his tombstone, and the cloud-covered peaks of Whiteface and Marcy to the north and south, towering five thousand feet above his head. By all means stop here a day. It will better you to stand a few moments over John Brown's grave, to enter the house he built, to see the fields he and his heroic boys cleared, the fences they erected and others standing incomplete as they left

FIGURE 5. Seneca Ray Stoddard, *John Brown's Grave and the "Big Rock," North Elba*, ca. 1896. Library of Congress, Prints & Photographs Division, LC-USZ62-107590.

them when they started for Harper's Ferry. What memories, if you are an American, will throng into your head as you stand beside that mound and traverse those fields! You will continue your journey a better man or purer woman from even so brief a visit to the grave of one whose name is and will ever be a synonyme [sic] of liberty and justice throughout the world.[125]

The risk of a definition fully formed is that, if one isn't careful, it can entomb a living history of movements, ideas, and people in the past tense, relegating things that happened long ago to a dead time that no longer matters rather than reanimating that history for the sake of those, today, who dream desperately of a new world.

9

Brown was still violently alive in 1858 when the tourists, like Emerson, began seasonally flooding the Adirondacks, looking "to kill time and escape from the daily groove," as Emerson's companion, the artist W. J. Stillman, put it.[126] But though the history of race had colored the Adirondacks from the moment the geologist Emmons named them, most white vacationers paid little attention to this past—except, perhaps, when it could be transmuted into the thrilling notion of a noble American Indian savagery, or when, after the war, one visited a tamed Brown safely in his grave. Brown himself, with his head full of apocalyptic dreams tinted red, wasn't all that different from many other wilderness pilgrims, who also fled their homes in search of a purer land. "I hoped here to find new subjects for art, spiritual freedom, and a closer contact with the spiritual world—something beyond the material existence," wrote Stillman, a longtime lover of the Adirondack woods.[127] He was tired of life in the city; he wanted something real, something transcendent, and the thing that impressed him most about the Great Northern Wilderness was its authenticity: "We found it in the Adirondacks: disguises were soon dropped, and one saw the real character of his comrades as it was impossible to see them in society. Conventions faded out, masks became transparent, and for good or for ill the man stood naked before the questioning eye—pure personality."[128]

Yet it can be hard to credit critique as radical when it comes from the ranks of privilege. Whatever discontent touched Stillman paled in comparison to the savagery daily beaten into black backs or the gear-grinding drudgery of modern life as a factory operative. But hearts break, even amid plenty. And one of the most transparently broken was J. T. Headley's, whose *The Adirondack; or, Life in the Woods* (1849) is an odd and influential early chronicle of Adirondack adventures. Neither a bundle of stories nor a field guide with sections on how to get where, it's something much more intimate: an epistolary collection bound in leather, each letter longingly addressed to "Dear H——," H. J. Raymond, the founder of the *New York Times* and the man to whom Headley dedicated the entire book—a dedication barbed with passive aggression, the sort of thing a jilted lover writes to his muse.[129] "Though you failed to accompany me in my trip to the Adirondack Region," Headley wrote, "yet I often thought of you in my long marches and lonely bivouacs."[130] He was isolated in the wilderness, and the Adirondacks felt strange and doubly lonesome; but after "the din and struggle of Broadway and Wall," Headley was in need of re-creation. The clangor, the flashing symbols and heated exchanges of New York, had brought on an "attack of the brain," had forced him to flee the "haunts of men" for the woods, and when he did, he seemed to find the freedom to finally live as his true self.¹³¹ "In the woods, the mask that society compels one to wear is cast aside, and the restraints which the thousand eyes and reckless tongues about him fasten on the heart, are thrown off, and the soul rejoices in its liberty and again becomes a child in action. . . . In wilderness there is no formality in the expression of one's feelings."¹³²

Headley's attack on the brain came from being "cheated, exasperated, slandered, and mortified," from having part of himself stolen away by a thieving society.¹³³ But here in the woods he bore witness to something new. To "Dear H." he poured out his thoughts, almost babbling in effusion:

> I love nature and all things as God has made them. I love the freedom of the wilderness and the absence of conventional forms there. . . . I love it, and I know it is better for me than the thronged city, aye, better for soul and body both. How is it that even good men have come to think so little of nature, as if to love her and seek her haunts and companionship were a waste of time . . . ? A single tree standing alone, and waving all day long its green crown in the summer wind, is to me fuller of meaning and instruction than the crowded mart or gorgeously built town.¹³⁴

Critique and freedom — these are the twinned attributes of wilderness that Stillman discovered when "in the silence of those nights in the forest, the whisperings of the night wind through the trees forced meanings on the expecting ear. I came to hear voices in the air."¹³⁵ They're the same whispered secrets that opened themselves to Headley as he sat under his lone witness tree in the late 1840s, at the very same moment when the eutopian agrarians just a few miles away in Timbuctoo, Blacksville, and Freeman's Home, were busy remaking the world.

Eight years later, S. H. Hammond published his *Wild Northern Scenes*, a wistful meditation on wilderness and society that sharpens Headley's heartache into a pointed critique. Like Headley, who wrote that "there is one kind of forest music I love best of all — it is the sound of wind amid the trees," sound is key to Hammond's experience of the wilderness, and he critically contrasts "the clank of machinery, the rumbling of carriages, the roar of the escape pipe; the scream of the steam whistle; the tramp, tramp of moving thousands on the stone sidewalk," against the "clear and musical and shrill" call of the loon, the "partridge drumming upon his log," the owl's "almost human haloo," the catbird, and brown thrush, and chervink, chickadee, wood robin, blue jay, wood sparrow—"a hundred other nameless birds that live and build their nests and sing among these old woods."[136]

Hammond heard the two places—the wild and the city—differently, and he also found they thought differently. The climax of *Wild Northern Scenes* is a staged conversation between Hammond's companions, in which one friend, Spalding, the great defender of Progress, crows: "'Everywhere, in all the departments of science, in every branch of the arts, improvement, progress has been going on with a sublimity of achievement unknown in any age of the past. . . . There is hope for the world in all this mighty progress. . . . Who will venture to assert where the limit to this progress may be found?'"[137]

Against such utopian faith, Hammond has his friend, the Doctor, reply:

> The good time of which you speak . . . will never come . . . The excesses of the world are a much more fruitful source of disease and death than the attritions of age. There is a constant struggle of nature to build up and beautify, to strengthen and recuperate, against the result of human excesses. . . . The outrages perpetrated upon nature by the conventionalities of the world alone, would be an insurmountable barrier to the realization of your idea. . . . It is a part of our civilization, an offshoot of the very progress of which you speak, a sort of necessity in practical results, at least, that men *shall* so live as to wage war against nature, and against themselves.[138]

One could argue that capitalism produces an endless stream of "goods." But pointing out, as the Doctor does, that what capitalism really produces is endless desire, and therefore endless scarcity (as well as an endlessly growing rag heap of yesterday's always-outmoded fashion), and that the manufactured excess of scarcity justifies continual war "against nature, and against [humans] themselves"—this seems closer to the truth.[139]

And yet, it isn't hard to wave Headley and Hammond and those like them away as privileged and therapeutic antimoderns out for a nice summer ramble in the woods. After all, their accounts helped turn the Great Northern Wilderness into a premier vacation spot for the middling and well-to-do: in the 1860s hotels started colonizing the Great Northern Wilderness, first around the prominent lakes, and then, especially in the 1870s, throughout the backwoods. By 1876 twenty bellied up to Lake George alone, and in the 1880s, driven in part by the coming of the iron horse, the phenomenon of the Great Camps was roaring along in high gear: dozens of elaborate, multiple-storied "rustic" structures, complete with ballrooms and fine china, went up in the woods, and the region became the pleasure ground for those titans of industry, the Morgans, Vanderbilts, and Huntingtons. The middling classes, too, found inns and camps aplenty, and by the closing decades of the nineteenth century, many wilderness lovers were worrying that the Adirondacks were proving too popular, worrying that the region was losing its distinction: lakes were crowded, trails eroded, fish and game depleted, and people, people, people were everywhere.[140]

But it's too easy to assume that the equation of wilderness with critique and freedom was trampled under the soles of well-heeled financiers and the pitter of a scrambling middle class; though neither Headley, nor Hammond, nor any other white Adirondack travel writer turned his critical eye to the social and economic sources of his own wealth, the story that by the end of the nineteenth century any incipient radicalism had been gentrified into a species of consumption is still too pat a plot line.

In fact, beginning in 1873, the Adirondacks had become famous as a place to cure the human body of the most feared by-product of industrialized city living—tuberculosis—thanks to Dr. Edward Livingston Trudeau. He had been inspired by Murray's prophecy in *Adventures in the Wilderness* that "the spruce, hemlock, balsam, and pine . . . yield upon the air . . . all their curative qualities," and he had turned to the Adirondacks to have the piney breezes purge TB from his own lungs, which were hemorrhaging badly.[141] It was a last-chance choice, and it worked. When he realized what had happened, he devoted the rest of his life to fighting tuberculosis with what was sometimes called the wilderness cure, and the Adirondack woods found itself a home to increasing numbers of hacking invalids as the nineteenth century came to an end.[142]

One among them was a young newspaper clerk from New York City whose tearing cough quickly ruptured into life-threatening illness.[143] Though the clerk had wasted away to a shuck of a human by the time he arrived in the Adirondacks, within a few months he could boast of his vitality: "The thermometer is close to zero. The air is crisp and cold. It might freeze your dainty city ears, but it is nothing to the hardy backwoodsman. Nothing to the young man."[144] The key to the wilderness cure was good living, and it began with the first breath. "There is no special atmosphere manufactured for house use," the clerk wrote, meaning that we cannot invent ourselves away from air pollution. "With no noxious odors, no defective drains or gas-pipes, no miserable furnaces, no double windows to shut out the oxygen," the wilderness simply was more supportive of life, and the young clerk drove the point home by contrasting a sallow, fetid industrial landscape against the pulmonary health of "cheery wood fires, open chimney-places, and a surrounding atmosphere of absolute purity."[145]

It's a vision of a different—of a better—world, that the white Adirondack tourists caught glimpses of, though it lacked the audacious eutopian commitment of Timbuctoo. But the Adirondacks are a big place, big enough for radicalism and reform both, and, in the end, I think the tourists and their pioneer neighbors dreamed a few Adirondack dreams in common. After all, the dying newspaper clerk, a scrivening Bartleby, discovered for himself that the right kind of wilderness could heal the lesions of consumption.

10

"If by definition wilderness leaves no place for human beings, save perhaps as contemplative sojourners enjoying their leisurely reverie in God's natural cathedral," wrote Cronon as the authoritative voice of the wilderness critics, "then also by definition it can offer no solution to the environmental and other problems that confront us."[146] Yet, if wilderness, at least the Great Northern Wilderness, bucks Cronon's attempts at definition, if some nineteenth-century wildernesses explicitly were a home for human beings and their eutopian societies, their intellectual histories of liberty and community, their traditions of critique and healing, then perhaps some wildernesses can begin to offer solutions for a wide range of environmental and social problems.[147]

A word can contain a landscape; change its definition, and you change the world. It's worth it to listen to the world speak back through the words used to describe it.

What made the Adirondacks a wilderness? They sounded like one: McCune Smith's and Frederick Douglass's woods ringing with freedom's ax strokes; Headley's favorite music, "the sound of wind amid the trees"; Stillman and his "whispering of the night wind through the trees"; Hammond's anguished question, "Where shall we go to find the woods, the wild things, the old forests, and hear the sounds which belong to nature in its primeval state?" Or the poet Alfred B. Street's sibilant breezes: "The soft Southwest says, Take thy rest / To-day upon nature's kindly breast!" Or W. H. H. Murray's Emersonian faith: "So with God: in the silence of the woods the soul apprehends him instinctively. He is everywhere. In the fir and pine, which, like the tree of life, shed their leaves every month, and are forever green."[148]

The Adirondacks were a wilderness because they smelled like one: the "balsamic odors" infusing the air of poet Homer Sweet's epic *Twilight Hours in the Adirondacks: The Daily Doings and Several Sayings of Seven Sober, Social, Scientific Students in the Great Wilderness of Northern New York* (1870). And the newspaper clerk's anticonsumptive spaces where "the proximity of pine and balsam trees is a most desirable thing." Or H. Perry Smith, who eagerly "snuffed the scented breezes" in *The Modern Babes in the Woods* (1872).[149]

The Adirondacks were a wilderness because they looked like one— and it's the look that rooted sound and smell into the single physical characteristic which, more than any other, defined nineteenth-century wilderness: trees.

A wilderness was a place of trees. A community of trees vast and unbroken and unsettled and untouched.

It's especially here, though, that the current critique of wilderness leads us astray: today, unbroken and unsettled and untouched are three forbidden words, to be handled ironically with scare quotes, deconstructed, and shown to be semantically adrift. Yet all three terms also have a history, and in the nineteenth century they weren't necessarily clothed in their current meaning. All three referred to agriculture's mark, and if one could have flown over the landscape, settled, broken, and touched land would have unrolled before the bird's eye as a continuous tapestry of farm fields, with little pockets of woodlots dappling the scene. Land that was unsettled, or untouched, or unbroken didn't mean that it was empty or unpeopled, or even uncleared—it just meant that the predominant view was of forest, a place where farm fields were periodic pauses in an otherwise uninterrupted sylvan narrative.[150]

This is one of the most surprising things about the Great Northern Wilderness: nearly every travel narrative and guidebook points here and there, begging its reader to consider the trees and mountains and rivers, as well as the towns and industries and workers and settlements. There were scores of farms, and more than two hundred mining outfits boring into the region's mountains—it was iron from the Great Northern Wilderness that ironically sheathed the ironclad hull of the Confederacy's gunboat *Monitor*, and Adirondack ore that was spun into the cables of the Brooklyn Bridge.[151] And of course, there was logging. Abundant raw materials gave rise to industry, and in the Great Northern Wilderness one could find work in twine, wire, or iron factories. Then there was the emergent tourist industry, the inns and innkeepers, the guides; Seneca Ray Stoddard's *The Adirondacks Illustrated* (1881) lists 146 different places to find a room "from the well-appointed hotel on the border to the rude log-house and open camp of the interior."[152] All of this in the Great Northern Wilderness, which, despite the human activity humming beneath its branches, was yet different. It was unsettled, untouched, unbroken—a wilderness.

And this untouched land felt different to a citified human body: it sounded and smelled and looked different, all of which helped breed a cultural understanding of the Great Northern Wilderness that emphasized its difference from the capitalistic world of chattel slavery and exploited factory labor. Unfortunately, some of the most influential critiques of wilderness have tended to code difference as negative—difference, after all, is the basis of exclusion—and instead have collapsed difference altogether into a sort of ecological melting pot in which a tree is simply a tree, whether it was planted by a squirrel in the wilds or by human hands in a suburban, upper-middle-class backyard.[153] It's an ecological and political ethic that values conformity and shared values rather than defending difference. But many in the nineteenth century understood that, though wild and urban were different, they weren't exclusively so, and visitors frequently found themselves bridging the distance from city to woods: Emerson did it when he imagined the intercontinental telegraph, as did Headley when he wrote, "how strange it seems to behold men thus occupied—living contentedly fifty miles from post office or village—and hear their inquiries about the war with Mexico," and as did McCune Smith when he, cosmopolitan sophisticate, found himself born again under forest trees after the death of his daughter.[154]

Recognizing difference means that many wilderness travelers thought of the woods not as empty, virginal, and waiting to be penetrated by culture—the definition of wilderness that prevails today—but instead as full: full of sound, full of smell, full of health, full of trees and animals. Full of life. It was a full place of wild where humans, privileged white tourist and radical black pioneer alike, could let down their guard and connect. Difference wasn't a lack, it was just different. If modern civilization had been disenchanted, the woods were different because they were yet a place where living things, some of them human, gathered, a place of mystery, a place of potential, a place of grace.[155]

When African Americans and abolitionists, invalids and tourists, came to the Great Northern Wilderness, they found critique—the witnessing forest forced consideration of the gap between the world they inhabited and the one in which they wanted to live—but they also found a positive social vision of what the world, their world, could be, a place of difference, a place of contact, a place for community, a place that looked, perhaps, like Thomas Cole's Adirondack painting *Schroon Lake* (ca. 1846), where improvement meant picturesque restraint, where

FIGURE 6. Thomas Cole, *Schroon Lake*, ca. 1846. The Adirondack Museum, Photography by Richard Walker.

life—human and sylvan—flourished and green shoots sprang from tree stumps. It's a theme that seemed to obsess Cole, who had spent much of the mid-1830s on his masterpiece series, *The Course of Empire*, an allegorical tale that traces human (and many thought American) history through five enormous paintings, from *The Savage State* (1834), to *The Arcadian or Pastoral State* (1834), the urban *Consummation of Empire* (1836) and internecine *Destruction* (1836), to finally find repose in *Desolation* (1836), where crumbling architecture is the only trace of humanity. In the background of each painting looms the same landmark mountain, anchoring each scene in the same deep historical landscape.

FIGURE 7. Thomas Cole. *The Course of Empire: The Arcadian or Pastoral State*, 1836. Oil on canvas, 39¼ × 63¼ in.; negative 6046; image 1858.2. Photography © New-York Historical Society.

Commentators have been arguing for almost two hundred years over whether Cole's series casts history as an inevitable cycle of rise and fall, or if he posed a warning to his viewers: perhaps humans could pause in arcadia, and avoid the slide from empire to emptiness. The thought stuck with Cole, and his peculiar landmark mountain reappears in *Home in the Woods* (1847)—and again in *Schroon Lake*. Perhaps, Cole's mountain seems to say, one could find the ideal wild civilization in the Great Northern Wilderness, the sort of place where the heartsick Headley, missing his Dear H., could write "the laws of Nature and Heaven are such that he who accumulates to live a life of idleness is made as miserable as the man he impoverishes in order to do it."[156]

11

I first set foot in the Adirondacks when I was sixteen, on July 4, 1996, 151 years to the day after Thoreau went to the Walden woods, thinking I followed in his footsteps by heading for the wilderness. I was backpacking with my best friend, and we left from the Garden parking lot, in Keene Valley, not far from where, back in 1849, James H. Henderson had given a letter to the local postman for Henry Highland Garnet, return address West Keene, "Timbucto"—a place of which I had never heard. We left our car, strapped on our gear, and set off into our own Edenic three-day venture up Mt. Marcy. We got soaked by rain, covered in black Adirondack mud, blistered, and lectured by a backcountry ranger on proper camping etiquette. But what I'll remember for the rest of my life is scrambling up Marcy's sides, on a trail that I seem to recollect the guidebook describing as "a waterfall of stones," while the balsam fir shrunk in size and huddled closer together until they seemingly turned into lichens. For the first time in my life, I was standing above the line of trees absolutely alone save for the company of one of the closest comrades I've ever known.

On that day, I caught a fleeting sight of something that has kept me coming back ever since on foot, skis, snowshoes, bike, and with climbing rope in hand. It's something that I chased around the country to tall mountains, narrow canyons, and deep woods, though I never could catch it. This same ineffable something drew me to my initial round of Timbuctoo research in 2000, and again in 2006, and yet again in 2011 and 2012. Chasing it eventually led to graduate school, where, on a research trip to the Adirondack Museum, I discovered that though the Adirondack landscape has stood witness to many watershed moments in my life, deep friendships forged and renewed, some of the starriest nights I've ever seen, early dates with my future wife, a nearly fatal climbing adventure, I knew next to nothing about the wilderness through which I had adventured. It turns out that the Adirondacks had only ever been for me a hard-edged mirror in which I discovered myself.

I suppose I should have known better, and I had inklings of their human history. In my senior year of college, I mentored a sweet though troubled sixth-grade boy from Moriah Central School, in Essex County, along the route that some of New York's pioneers would have taken to find their lands. It's a beautiful area, and great for hiking, but poor—indeed, upstate New York is one of the state's poorest, most ignored regions (because the population is slight, the area draws little interest from the political class and even less from the economic elite). Essex and Franklin Counties respectively rank forty-ninth and fifty-sixth out of sixty-three New York counties in terms of median income.[157] My student and I talked about skateboarding, and playing in bands, and we swapped punk-rock mix tapes, but there was always a dark streak: I heard tales of neglect and alcoholism and hardship, tales of unhappy homes there in the forever-wild Adirondack wilderness. I once tried to turn the tide of conversation by asking him and his friends if they spent much time hiking or skiing, seeking salvation in the empty wilds, but I was met, instead, with silent stares, blank and incredulous and still too innocent to be resentful, that I now understand spoke eloquently of my relative privilege and blinkered naïveté. At the time, I didn't quite get it.

Five years later, in the summer of 2007, on the way to the trailhead of what turned out to be an aborted 130-mile hike, I saw a sticker slapped on a storefront, and though I wish I could say that I finally came to understand something about the wilderness after reading one of the great works of environmental history, it was that bumper sticker that finally crumbled the definition of wilderness around which I had made my life orbit:

It's no damned PARK
It's the ADIRONDACKS
It's our HOME
It's where we WORK

There's a part of me that loves the sheer orneriness of that sticker, for I am, after all, a country boy myself, from rural upstate New York, no less, and I grew up with bumper stickers like these ("If they call it tourist season, why can't we shoot them?" is a popular one), in a world with one all-governing typological distinction: there are locals and there are city people, and city people are wealthy folks who come from somewhere to play while everyone else sweats for their pleasure. Sometimes city folks dismiss these stickers as backward regional chauvinism, or even worse, a dangerously explicit exclusion, and they're right to do so; but stickers such as these are also cries for visitors to pay attention not only to the scenery but to their own imported cultural stereotypes. We who live in the country matter, and there's a poetic rhythm to that Adirondack sticker born of repetition and half rhyme, as well as a point: *Adirondacks, home, work,* and *park*—the only word with a negative connotation, the only interloper, for, of the Park's more than six million acres, by far the largest nature reserve in the United States (larger than the entire state of Massachusetts), only 44 percent as of 2014 is actually protected land, most of it wilderness, owned by the state. The rest is private land on which you'll find ski resorts, logging companies, amusement parks, car dealerships, a few colleges, gas stations, pizza joints, bars, a prison, an Olympic training complex, and thousands of homes.[158]

Yet the Adirondack Park isn't going anywhere. Written into the state's constitution of 1894 is the famous "Forever Wild" clause, article 7, section 7, which reads: "The lands of the State, now owned or hereafter acquired, constituting the Forest Preserve as now fixed by law, shall be forever kept as wild forest lands. They shall not be leased, sold or exchanged, or be taken by any corporation, public or private, nor shall the timber thereon be sold, removed or destroyed."[159] The land had previously been protected as a forest preserve under an 1885 law, but the State Forestry Commission and the foresters charged with guarding the forests were regarded by nearly everyone as inept, negligent, and perhaps in cahoots with the fly-by-night logging operations—the true bark eaters—that were torturing the timber and stripping the land bare. The park—which included the forest preserve—was created in 1892 as a last-ditch effort to curb profit-oriented greed.[160] It also buried older notions of inhabited, working wildernesses, instituting, for the first time in that landscape's history, a definition that finally fits today's notion of the empty wilderness.

No person can cut a stick of timber, divert a stream, or dig a shovelful of ore in the preserve without a constitutional amendment—the Adirondacks, or at least the parts of them owned wholly by the state, have been protected as a space for play, it seems, indefinitely. Yet it doesn't take much more than an hour's time with any of the key works of Adirondack history to discover that hunting and fishing and camping, though an important aspect of Adirondack preservation, were all secondary interests. In fact, beginning in the mid-nineteenth century, the Adirondacks were considered one of the Empire State's most valuable landscapes because the intact forest cover was testimony to the state's economic might.[161] George Perkins Marsh's widely read *Man and Nature* (1864) forcefully drove home the connection between cleared land and decreased river flow, and for a state such as New York, whose wealth depended on the navigability of the Erie Canal and Hudson River, for a state whose growing mass of Manhattanites needed clean water, preserving the Adirondacks meant ensuring financial fluidity. Preserving the city meant preserving the wilderness, and so even the modern, negative definition of Adirondack wilderness as a place empty of all civilization has always been yoked to one of the most urban spots on earth.[162]

It's all of our loss that the Adirondacks' remembered past has been stripped of its countermodernity, of its living potential for a radically better world. Eight years after writing *Invisible Man* (1952), one of the twentieth century's great meditations on race, space, and history, Ralph Ellison argued that "if we don't know *where* we are, we have little chance of knowing *who* we are, that if we confuse the *time* we confuse the *place*; and that when we confuse these we endanger our humanity, both physically and morally."[163] History, Ellison knew, is made in the present from the remains of time long gone. History helps us define who we are and our place in the world—and that's exactly what we've long used trees for, as well. As it turns out, the word *tree* is related to the word *truth*: they once diverged from the same root definition.[164]

I'm not sure there's a lesson to be found in the Adirondack wilderness, other than that it's risky to pin a living definition in place, risky to read the present back onto whatever came before. Or perhaps the lesson is simply, and humbly, to listen: listen to the trees.

When Jermain Wesley Loguen wrote to James McCune Smith of his Adirondack prospects, he noted that some locals were taking advantage of the disoriented newcomers. But Loguen didn't worry himself too much; he knew that you can't erase truth: "Around every tree on which the [surveyor's] figures were engrained, there will be found trees, called witnesses, that are blazed. . . . In this way . . . one may know whether he has arrived at a spot where a landmark was fixed by the surveyor . . . if he finds the witnesses."[165]

The trees remember. We just need to find the right words.

INTERMISSION

I watched the trees slowly recede, waving their despairing arms, seeming to say to me: "What you fail to learn from us to-day, you will never know. If you allow us to drop back into the hollow of this road from which we sought to raise ourselves up to you, a whole part of yourself which we were bringing to you will vanish forever into thin air."

MARCEL PROUST, *Within a Budding Grove*[1]

W. H. H. Murray's *Adventures in the Wilderness* (1869) was a sensation. Upon its publication, a stampede of Civil War–weary tourists, the leading edge of a great seasonal migration that would come to be known as "Murray's Fools," lit out for the Adirondack woods, many of them looking for the wilderness tonic, cleanly innocent of the past and all its burdens, that Murray had promised. Murray was one of the early, influential hawkers of Adirondack emptiness, and he tried his best to turn the Adirondacks into a raw, before-the-white-man landscape of (not too much) danger, (just enough) physical hardship, and (if not for the loggers) endless sylvan plenty.

Yet *Adventures in the Wilderness* is weird, and its final chapter, "A Ride with a Mad Horse in a Freight-Car," unsteadily walks the plot lines

FIGURE 8. Arthur Fitzwilliam Tait, *A Good Time Coming*, 1862. The Adirondack Museum, Photography by Richard Walker.

of an unresolved story-within-a-story—a chapter more postmodern literary hallucination written a century too soon than an earnest advertisement for wilderness travel, a conclusion that makes you wonder, once the book is done, did I miss the point?

The book was published in 1869, the same space-and-time-annihilating year that a golden spike nailed together the iron rails of the transcontinental railroad somewhere out on a treeless flat in Utah's Great Basin. This chapter of Murray's recounts a story he heard when camping with a friend on the banks of Raquette Lake in July 1868, and he took it down when into his firelight stepped the Stranger—silent, ghostly, possessed of a wartime tale that began in 1862:

> It was at the first battle of Malvern Hill,—a battle where the carnage was more frightful, as it seems to me, than in any this side of the Alleghenies during the whole war,—that my story must begin.... About 2 P.M., we had been sent out to skirmish along the edge of the wood in which, as our generals suspected, the Rebs lay massing.... We had barely entered the underbrush when we met the heavy formations of Magruder in the

very act of charging. . . . They were on us and over us before we could get out of the way. . . . When the last line of Rebs had passed over me, I was left amid the bushes with the breath nearly trampled out of me, and an ugly bayonet-gash through my thigh; and mighty little consolation was it for me at that moment to see the fellow who run me through lying stark dead at my side, with a bullet-hole in his head, his shock of coarse black hair matted with blood, and his stony eyes looking into mine. . . . Never have I seen, no, not in that three days' desperate mêlée at the Wilderness, nor at that terrific repulse we had at Cold Harbor, such absolute slaughter as I saw that afternoon on the green slope of Malvern Hill. The guns of the entire army were massed on the crest, and thirty thousand of our infantry lay, musket in hand, in front. For eight hundred yards the hill sank in easy declension to the wood, and across the smooth expanse the Rebs must charge to reach our lines. It was nothing short of downright insanity to order men to charge that hill; and so his generals told Lee, but he would not listen to reason that day, and so he sent regiment after regiment, and brigade after brigade, and division after division, to certain death. . . .

It was at the close of the second charge, when the yelling mass reeled back from before the blaze of those sixty guns and thirty thousand rifles . . . that I saw from the spot where I lay a riderless horse break out of the confused and flying mass, and, with mane and tail erect and spreading nostril, come dashing obliquely down the slope. Over fallen steeds and heaps of the dead she leaped with a motion as airy as that of the flying fox, when, fresh and unjaded, he leads away from the hounds, whose sudden cry has broken him off from hunting mice amid the bogs of the meadow. So this riderless horse came vaulting along. . . . When I saw this horse, with action so free and motion so graceful, amid that storm of bullets, my heart involuntarily went out to her, and my feelings rose higher and higher at every leap she took amid the whirlwind of fire and lead. And as she plunged at last over a little hillock out of range and came careening toward me as only a riderless horse might come . . . I forgot my wound and all the wild roar of battle, and, lifting myself involuntarily to a sitting posture as she swept grandly by, gave her a ringing cheer.

FIGURE 9. Portrait of Pvt. Edwin Francis Jemison, 2nd Louisiana Regiment, C.S.A., killed at the battle of Malvern Hill, ca. 1862, rephotographed 1961. Library of Congress, Prints & Photographs Division, LC-B8184-10037.

This wild mare, dripping the blood of its former owner, approached the Stranger, and she was the most beautiful thing he had ever seen. Murray devotes two pages to the Stranger's loving, sensuous description of every chocolate body curve and glinting tress of hair, every personality trait—passages of true love revealed amid apocalypse.

When the Stranger's men found him later, wan and nearly insensible from shock and blood loss, the horse refused to leave his side and accompanied him back to Washington, where he recovered from his injuries. Upon regaining his senses, he awoke to find the horse tending his bedside. Breaking down in tears of gratitude, the Stranger named her Gulnare, after Jullanár—the beautiful slave girl of *The Arabian Nights' Entertainments*, sold to King Sháh-Zemán for ten thousand pieces of gold—a girl whose beauty "was such as would cure the malady of the sick, and extinguish the fire of the thirsty."[2] The Stranger loved his equine Gulnare:

> I, so far as man might be, was hers. . . . I am not ashamed to say that I put both my arms around her neck, and, burying my face in her silken mane, kissed her again and again. Wounded, weak, and away from home, with only strangers to wait upon me, and scant service at that, the affection of this lovely creature for me, so tender and touching, seemed almost human, and my heart went out to her beyond any power of expression, as to the only being, of all the thousands around me, who thought of me and loved me.

FIGURE 10. Emanuel Gottlieb Leutz, *Westward the Course of Empire Takes Its Way* (Mural Study, US Capitol), 1861. Oil on canvas, 33¼ × 43⅜ in. (84.5 × 110.1 cm). Smithsonian American Art Museum, Washington, DC / Art Resource, NY.

Healthy again, The Stranger and Gulnare returned to the front for three more terrible years, and, lucky enough to have survived together, paraded at war's end through both Richmond and Washington in triumphant, relieved celebration, a man and his love: "And then I thought of home, unvisited for four long years,—that home I left as a stripling, but to which I was returning a bronzed, brawny man."

That night they slept together for want of lodging, and, once daylight broke over the war-ravaged landscape, boarded their final train, sharing a car, man and horse dreaming of the good times sure to come. The Stranger fell asleep. But when he opened his bright eyes he noticed that Gulnare's were dull and heavy.

> Never before had I seen the light go out of them. The rocking of the car as it went jumping and vibrating along seemed to irritate her. She began to rub her head against the side of the car. Touching it, I found that the skin over the brain was hot as fire. Her breathing grew rapidly louder and louder. Each breath was drawn with a kind of gasping effort. The lids with their silken fringe drooped wearily over the lustreless eyes. The head sank lower and lower, until the nose almost touched the floor. The ears, naturally so lively and erect, hung limp and widely apart. The body was cold and senseless. A pinch elicited no motion. Even my voice was at last unheeded. To word and touch there came for the first time in all our intercourse, no response. I knew as the symptoms spread what was the matter. The signs all bore one way. She was in the first stages of phrenitis, or inflammation of the brain. In other words, *my beautiful mare was going mad.*

The cure was to bleed her. But The Stranger had misplaced his pocketknife.

> "My God!" I exclaimed in despair . . . "must I see you die, Gulnare, when the opening of a vein would save you? Have you borne me, my pet, through all these years of peril, the icy chill of winter, the heat and torment of summer, and all the thronging dangers of a hundred bloody battles, only to die torn by fierce agonies, when so near a peaceful home?"

But little time was given me to mourn. My life was soon to be in peril, and I must summon up the utmost power of eye and limb to escape the violence of my frenzied mare. Did you ever see a mad horse when his madness is on him? Take your stand with me in that car, and you shall see what suffering a dumb creature can endure before it dies. In no malady does a horse suffer more than in phrenitis. . . . A horse laboring under an attack of phrenitis is as violent as a horse can be. . . . He is unconscious in his violence. He sees and recognizes no one. There is no method or purpose in his madness. He kills without knowing it. . . .

I knew what was coming. . . .

I took my position in front of my horse, watchful and ready to spring. Suddenly her lids, which had been closed, came open with a snap, as if an electric shock had passed through her, and the eyes, wild in their brightness, stared directly at me. And what eyes they were! The membrane grew red and redder, until it was of the color of blood, standing out in frightful contrast to the cornea. The pupil gradually dilated until it seemed about to burst out of the socket. . . . Spasms . . . ran through her frame. . . . Then followed exhibitions of pain which I pray God I may never see again. . . .

The mare raised herself until her shoulders touched the roof, then dashed her body upon the floor with such a violence which threatened the stout frame beneath her. I leaned, panting and exhausted, against the side of the car. Gulnare did not stir. She lay motionless, her breath coming and going in lessening respirations. I tottered toward her, and, as I stood above her, my ear detected a low gurgling sound. Gulnare, in her frenzied violence, had broken a blood-vessel, and was bleeding internally. Pain and life were passing away together. I knelt down by her side. I laid my head upon her shoulder and sobbed aloud. Her body moved a little beneath me. I crawled forward and lifted her beautiful head into my lap. . . . I smoothed the tangled masses of her mane. I wiped, with a fragment of my coat, torn in the struggle, the blood which oozed from her nostril. I called her by name. My desire was granted. In a moment Gulnare opened her eyes. The redness of frenzy had passed out of them. She saw and recognized me. I spoke again. Her eye lighted a moment with the old and intelligent look of love. Her ear moved; her nostril quivered gently as she strove to neigh. The effort was in vain. Her love was greater than her strength. She moved her head a little, as if she would be nearer me, looked once more with her clear eyes into my face, breathed a long breath, straightened her shapely limbs, and died. And there, holding the head of my dead mare in my lap, while the great warm tears fell one after another down my cheeks, I sat until the sun went down, the shadows darkened in the car, and night drew her mantle, colored like my grief, over the world.[3]

FIGURE 11. A. J. Russell, *Hall's Fill above Granite Canon*, ca. 1869. From A. J. Russell, *Photographs Taken during Construction of the Union Pacific Railroad.* Yale Collection of Western Americana, Beinecke Rare Book and Manuscript Library.

ACT THREE   REVELATOR'S PROGRESS

Sun Pictures of the Thousand-Mile Tree

"O.K., here we go." Roswell lit a ruby darkroom lamp. Took a dry plate from a carrying case. "Hold this a minute." Started measuring out liquids from two or three different bottles, keeping up a sort of patter meantime, hardly any of which Merle could follow . . . Stirring it all in a beaker, he put the plate in a developing tray and poured the mixture over it. "Now watch." And Merle saw the image appear. Come from nothing. Come in out of the pale Invisible, down into this otherwise explainable world, clearer than real. It happened to be the Newburgh asylum, with two or three inmates standing in the foreground, staring. Merle peered uneasily. Something was wrong with their faces. The whites of their eyes were dark gray. The sky behind the tall, jagged roofline was nearly black, windows that should have been light-colored were dark. As if light had been witched somehow into its opposite.

THOMAS PYNCHON, *Against the Day*[1]

1

Captain A. J. Russell, smelling of wood smoke and the balsam fir that made up his backcountry camp beds, had arrived, and it was time to work. First, he and his assistant unloaded the fussy old wagon and readied the just-thrown-together field laboratory for use: with great care Russell prepared the dark tent and into it, in their particular basins, beakers, and baths, went the various concoctions of silver, resin, water —

all the chemicals required for making a glass-plate photographic negative.[2] Then he removed his cameras. Their polished glass, brass, and wood playfully reflected the western light as he paced around the site, noting the wind coming out of the east, fretting about the desert-dry Utah air spiced by silver-green sage and burnt dust. He soon selected a view according to the light and the aspect of the surrounding hills — the composition slowly gaining form and structure, if only in his head. How to capture the evergreen, telegraph pole, and bleached tree skeleton? What to do with that hulking pile of rubble? Where to have the train stop? Should the hills hug the picture's sides or remain beyond the camera's monocular eye, invisible and unknown? He drew on his experience as a landscape painter, on his European studies, to imaginatively shift the raw space before him into something expressive, something more like the landscape visible to him only when he closed his eyes, or when he gazed off, unfixedly, into the middle distance.[3]

Gathering together his subjects and then dispersing them with precise directions on where to stand, sending one up to the very top of the tree, a small flag in his hand, and one to the apex of the ballast pile, Russell was ready to begin the alchemy of turning inanimate cyanide and gun cotton, silver and glass, into an image lifelike enough that this new art of photography raised fears as to the reality of nature. These sun drawings, so exact and so detailed, so mysteriously true to life, elicited fear at the same time that they commanded admiration, for though photographic technology seemed to breathe Progress, its soul remained resolutely premodern: a sort of necromancy, a way to freeze life and reanimate the dead, a ripple in space that brought the distant near to hand.

No one really knew how it worked. Even sober science stood silent when it came to explaining how silver and simple chemicals simulated life: "Light acts on the nitrates of silver. Why? No one knows and perhaps no one will ever know, but the fact is manifest," wrote one photography textbook author.[4] Where science failed, myth rushed in, and to some with a taste for the classics, the photographer was a latter-day Prometheus: "before another generation has passed away, it will be recognized that a new epoch in the history of human progress dates from the time when He who 'never but in uncreated light / dwelt from eternity' took a pencil from the hand of the 'angel standing in the sun,' and placed it in the hands of a mortal."[5] Some thought the intense focus of the photographic sitter, as she stared at the lens, burned her image into the glass plate. Others, that one's shadow literally stuck to the camera's film. Writers imagined the horrific, erotic possibilities of the photograph's ability to duplicate reality: What if the people in the images were actually alive? What if a photograph was a window into the private lives of its subjects?[6] The less gothically inspired wondered what potential for a new world, what potential for revelation, the new technology held. The popular dime-novel author Ned Buntline, in *Love at First Sight; or, The Daguerreotype, a Romantic Study in Real Life* (1848?), told the story of an image of a beautiful, destitute young woman, displayed in a Boston daguerreotypist's studio. When one day a wealthy patron saw it, he fell madly in love and searched the city for his never-seen-in-the-flesh love; ultimately he discovered her utterly penniless, wracked by illness, and one step away from prostituting herself on Boston's mean streets. But she's true to her image, and the patron rescued her from "this crooked, chequered, wilderness of a town, where poor girls are as plenty as flowers in the prairie grass."[7] It's a fantastic story, sure, but if living images could be made to stick to inanimate glass or metal, who could judge where the line between fantasy and reality lay?

Russell, acolyte of the sun, knew his ritual well: with great reverence, he first slipped a ten-by-thirteen-inch glass plate from its protective wooden box.[8] Having already placed his camera and tripod, arranged his subjects, and gingerly manipulated the camera's lens into a fine focus, he had only a very few minutes to transform the untreated piece of glass into a negative, which had to be exposed while the silver coating its surface was still tacky and damp. Not only did he have to fight time, but the wind and the general dryness of the southwest Rocky Mountains, too.[9] First, the plate had to be cleaned, preferably with Tripoli powder and alcohol. Next it was dried with Japanese paper—the slightest oily smear or wind-borne fleck of dust would ruin the bond between collodion and glass. Then, by the light of a red candle-powered lamp, the collodion itself had to be poured deftly on the plate in a very particular way so that the liquid flowed, evenly, over the entire surface, never touching the same bit of glass twice. It had to form a perfect union: a shaky hand or astigmatic coating led to distortion.

Sometimes the lunatic collodion, for whatever mysterious reason of its own, simply refused to cohere, and the photographer could do nothing but wait for it to change its mind.[10] But on that day in 1869 it chose to play along. Russell was able to prepare the glass properly, and then plunge the plate into a bath of mercurial silver nitrate before surely sliding the then-fertile negative into a frame called a dark slide, a device that shielded the negative from any stray sunbeam as he methodically hurried from dark tent to tripod, whereupon he slipped the whole works into the back of his expectant camera; the side of the dark slide facing the lens was outfitted with a hatch, and, checking over his view one final finicky time, he pulled this hatch from the frame, exposing the viscous collodion on the plate, hollered something like "keep quite still" at his anxious sitters, exhaled evenly, as if squeezing a rifle's trigger, to steady his hands—and then he removed the lens cap, letting in the day.

Then he waited.

He waited as the seconds dragged past. He waited long enough for the wind to blur the flag in the hand of the man at the top of the tree. He waited for perhaps a few minutes, not counting out the time mechanically, with a watch, but instead feeling the picture emerge as the camera's lens spilled the sun's rays onto his glass slide, judging by long experience when the outside world had been perfectly mirrored, when the inert plate, an idiosyncratic individual "exposed in accordance with the nature of the collodion," as the textbooks put it, came almost alive. And when he had waited long enough, he gingerly replaced the lens cap, shut the dark slide's trap, removed the plate—pregnant with possibility—and, carefully shielding it from the suddenly hostile sun's rays, returned to the safe red light of his dark tent to pour a sulfate of iron solution over the negative, to watch as the picture appeared "gradually as if by enchantment, clear, pure, sharp; the details . . . admirably distinct; the light . . . free from stains, and the blacks . . . represented by distinct tones varying according to the depth of shadow."[11] Russell's practiced eye would be able to translate the negative, with its perfect substitution

FIGURE 12. A. J. Russell, *1000 Mile Tree. 1000 Miles West of Omaha*, ca. 1869. From A. J. Russell, *Photographs Taken during Construction of the Union Pacific Railroad*. Yale Collection of Western Americana, Beinecke Rare Book and Manuscript Library.

of night for day, into what the final product would be: a richly gold-toned image of a beautiful mystery.[12]

2

Russell made his photograph of a witness tree located at what the railroad companies liked to think was the exact point that Union Pacific tracklayers hammered down their thousandth mile of iron rail, in Utah's Weber Canyon, in 1869. It's an albumen print, a print made with egg white, which traps the image-laden silver compound a shallow breath above the paper, an interval just wide enough for the image, in the case of *1000 Mile Tree*, to cast its own question-mark-shaped shadow forward in time to today.[13]

Part of the sheer Dionysian pleasure of the photograph lies in abandoning oneself to this living work of art, in letting one's eye be drawn by the strong, converging diagonal lines of the hillsides to the pile of ballast in the photograph's center, in recognizing all the suggestive, resonant variations on arboreal themes—the vertical lines of both the Thousand-Mile Tree and the two telegraph poles to its right; the horizontal, bleached tree skeleton in the immediate center foreground, which implicitly gestures to the stripped and trimmed tree trunks lying underneath the train's steel rails—in wondering, who are the fifty-two people, an inconsequential multitude, crowded around the base of the tree? In wondering after the absent train itself. Why do we only see the tender, the engine's tagged-on afterthought?[14]

What does *1000 Mile Tree* mean?

The plain of view is flat. There's little depth. Nothing to pull the eye out into a deep distance—the future's space. And because of this flatness, one's eyes are again led back to the center by those converging diagonal lines, back to the rubble pile, to the telegraph pole that juts from its peak, to the man who looks to be nailed to its cruciform shape, back to Golgotha, the hill on which Jesus was crucified, here transposed by Russell to Utah in order to make the Thousand-Mile Tree a witness to the sacrificial annihilation of space and time.

All of which is odd, since Russell was in Weber Canyon in 1869 only because, a year earlier, he had been hired by the Union Pacific Railroad to document the completion of its transcontinental line, which he did in what is one of the most famous American photographs ever taken.[15] Entitled *East and West Shaking Hands at Laying Last Rail*, it's the iconic view of the railroad's completion.[16] Locomotives from both the Union and Central Pacific Railroads, America's first megacorporations, meet cowcatcher to cowcatcher. Men on each engine joyously extend bottles of champagne to one another as the corporate chiefs Leland Stanford and Thomas Durant, flanked by a large crowd, stand in the center and shake hands—human mirrors of their machines. *East and West Shaking Hands* certainly captures the triumph of technology over space, and the promise of the nation's post–Civil War rebirth into a world where hard work would supposedly lead to economic well-being for everyone showed on the expectant faces of those gathered for the occasion.[17] It seems like technology and humanity are harmonious extensions of

FIGURE 13. A. J. Russell, *East and West Shaking Hands at Laying Last Rail*, ca. 1869. From A. J. Russell, *Photographs Taken during Construction of the Union Pacific Railroad*. Yale Collection of Western Americana, Beinecke Rare Book and Manuscript Library.

each other, each with a will to bridge the continent.[18] "The great Rail Road problem of the age has been solved," crowed Russell at the time.[19] He was from the small town of Nunda, in western New York, and had modern transportation in his blood: his family worked at canal and railroad building throughout the mid-nineteenth century, and he thrilled to the moment when the rails were joined.[20]

But it's hard to read such unalloyed enthusiasm into *1000 Mile Tree*. It's obviously neither a travel memento nor an advertisement: the passengers are too removed for any sense of vicarious identification, and even despite the fine-grained fidelity of the collodion, distance makes sure that there's little more than mere facts—the number of travelers, their dress—to be gleaned, no smiles or frozen laughs, no exchanged glances, no facial twitch of impatience or expectation, nothing that seems like it would sell. There's no human connection, either, although there are shades of real people who stand alienated from one another and from their surrounding world, a leprous distance that also marks them off from us, the audience, as frozen time flits past, all of us delivered to Golgotha by a steam engine and a camera, a scene presided over by the one immovable, permanent, historical figure: the Thousand-Mile Tree.

But why? What is it that *1000 Mile Tree* wants from us? What does it mean?

3

From its very beginning photography has been understood as both a utopian technology and a troubling new medium. "*Form is henceforth divorced from matter,*" announced Oliver Wendell Holmes just after photography's daybreak in 1859—and this division was part of photography's techno-utopian promise of ultimate objectivity.[21] No longer would one have to rely on a painter's hand in conveying the facts of the visible world; the image that the camera created seemed completely independent of human skill. Indeed, early photographers were often thought not to be artists—those lone geniuses who spent their nights and early mornings wrestling with an invisible muse—but technicians who merely manipulated a machine.[22] The hope of mechanical objectivity pervades the history of photography—even to this day—and in the 1840s and 1850s, a true photographic portrait was thought to capture an inner essence of the sitter: long exposure times supposedly allowed a person's true character to emerge.[23] Sometimes, early theorists thought, this true character revealed itself as a violent, antisocial one—this is the logic behind the infamous "rogue's gallery," in which photographs illustrating notorious physiognomic features were gathered into a sort of field guide, which, it was hoped, could be used not only to apprehend criminals on the run but also to predict those whose minds were bent toward violent crime.[24] The camera as a technology of social control found an influential adherent in Louis Agassiz, who felt that photographs captured divine truth. In 1850, sniffing what to him seemed to be the sour reek of social justice in the air, he commissioned J. T. Zealey of

Columbia, South Carolina, to photograph a number of Southern slaves. Agassiz hoped these images might prove that different races were created individually by separate thoughts of God, and therefore that race was an immutable characteristic, providing, once and for all, an irrefutable scientific basis for white supremacy.[25]

Yet this very modern celebration of defining one thing (or person) from another with an image caused a great deal of anxiety in the nineteenth century, precisely because the longed-for permanent division proved not quite up to holding back the human floodwaters of curiosity, sympathy, and connection. Photographic difference often encourages us to look over the wall in search of contact, only to find ourselves shaken by whomever, or whatever, looks back. The theorist Roland Barthes muses that there is something unsettling about being presented with a photograph of oneself that corresponds to no personal memory, like discovering a lost page from your own autobiography. There's something stranger still about being presented with an old photograph of an unfamiliar person, because we, the audience, are aware of a lapse in time of which the photograph's subject can only ever remain innocent. We may not know what the person's story is, but we know how it inevitably ends. For Barthes, photographs bring us face to face with the unrequited longing of an unremembered past and, ultimately, death.[26] Still, we look.

Even though photography and its preservation of difference might seem at odds with the railroad and its genius of standardization—one thinks of the inventions of the time zones, a corporate decision unilaterally made by the railroads in 1883—both the camera and the engine seemed to many nineteenth-century observers cast from the same mold: both relied on mysterious forces that nevertheless appeared pedestrian (photosensitive chemicals and thermodynamics, light and steam). Both were technological wonders. Both claimed to conquer nature. Edgar Allan Poe thought that the camera "must undoubtedly be regarded as the most important, and perhaps the most extraordinary triumph of modern science."[27] It's a sentiment that could be applied equally well to the railroads. And, as it turns out, the first plans for the transcontinental railroad were developed by a daguerreotypist.

When John Plumbe Jr., first arrived in Wisconsin in 1836, he had already served as an apprentice under the renowned civil engineer Wirt Robinson in surveying the nation's first set of interstate railroad tracks through the Allegheny Mountains. The sight of the treeless plains must have tripped a crucial circuit in his sensitive imagination, for soon after settling down, he began proselytizing for a transcontinental railroad and a future that he was convinced was imminent.[28] In 1837 and 1838, Plumbe petitioned Congress through his territorial representative, and in 1838 he actually secured a contract, and $2,000, to survey the route from Milwaukee to his newly adopted hometown of Sinipee, Wisconsin. Not everyone was as convinced as Plumbe that his plan was practical, or even possible, and his ideas were regarded by many as "'wild and visionary in the extreme—premature a century at least—the emanation of a mad enthusiast's brain.'"[29] When the territorial governor passed through town, his secretary had the opportunity to hear Plumbe expound on his idea and noted that "it was said by the villagers that the young man was crazy."[30]

As these things tended to do, Plumbe's dreams for a transcontinental railroad slid into his adventures in boosterism, and in 1839—the year that Louis-Jacques-Mandé Daguerre invented the device that would bear his name and play such a large role in Plumbe's future—he published a small tract entitled *Sketches of Iowa and Wisconsin, Taken during a Residence of Three Years in Those Territories*, which sang, in the most brash of melodies, the virtues of the Midwest. In many ways it's a conventional bit of advertisement for a speculative venture: he included dozens of letters from respectable people who claimed that they had never seen soil so black, weather so delightful, crops so plentiful. But Plumbe was only warming up. "A *National Rail Road*," he continued, "has already been commenced under the auspices of the General Government ... and an enlightened Congress has now entered upon a plan, whereby, ultimately, to secure to the United States, the *free* use, *forever*, of a Grand National Rail Road from lake Michigan to the Pacific Ocean!" (Neither claim was true.) Yet Plumbe was not simply a huckster; he seems to have truly believed that the transcontinental railroad was a common good: "The high, the low, the rich, the poor, will all be benefited—*none* will be injured," he stressed in more than twenty newspaper articles between 1847 and 1849.[31]

Perhaps it's fitting that a man so enraptured by technology would be equally captivated by the mechanical ability to preserve the present just before it slid into the past. And so, in 1840, when Plumbe found himself back in Boston, he took it upon himself to learn the year-old art of sun drawing from one of Daguerre's agents. Soon after, he began promoting himself as a "professor of photography" in a variety act that included a female magician, a head-reading phrenologist, a giant, and the Tattooed Man.[32] From the carnivalesque to the respectable was only a readjustment of focus, and within five years Professor Plumbe was the most recognized name in American photography.[33]

A young Walt Whitman visited Plumbe's New York studio in 1846 and came away struggling to find language to match the combination of modernity and superstition embodied by the photograph. The illusionary, doubling power of Plumbe's images worked their ironies on the journalist's prose, and Whitman became enchanted by his eyes' revelations: "You will see more *life* there—more variety, more human nature, more artistic beauty . . . than in any spot we know of." Plumbe had created an alternate city, awhirl in the stilled bustle and mute roar of the streets. "Whichever direction you turn your gaze, you see nought but human faces," Whitman sang. "There they stretch, from floor to ceiling—hundreds of them . . . human eyes gazing silently but fixedly upon you, and creating the impression of an immense Phantom concourse—speechless and motionless, but yet *realities*. You are indeed in a new world—a peopled world, though mute as the grave." It's something of a sinister vision, the daguerreotype that peeps back, stripping the powerless audience of its anonymity, and it thrilled Whitman, who was always up for a bit of transgression:

> There is always, to us, a strange fascination in portraits. . . . It is singular what a peculiar influence is possessed by the *eye* of a well painted miniature or portrait—It has a sort of magnetism . . . An electric chain seems to vibrate, as it were, between our brain, and him or her preserved there so well by the limner's cunning. Time, space, both are annihilated, and

we identify the semblance with the reality.—And even more than that. For the strange fascination of looking at the eyes of a portrait, sometimes goes beyond what comes from the real orbs themselves."[34]

Magnetism, spiritualism, the mysterious effects of electric forces as they pulsed through the ether: Whitman understood that these pseudoscientific things were all tied to the collapse of space and time and that Plumbe was a high priest who used the insights of science to create a shadow kingdom eerily similar to, and yet distinct from, the ordinary streets of New York.

But these heady days soon ended, and the professor, financially overextended, his photographic empire crumbling to dirt, headed to California in 1849, not as an Argonaut, but with the addition of California and the Mexican Cession making a transcontinental rail link seem not so foolish after all, on a quest to promote his steam-driven obsession. Along the way, he traded the honorific *professor* for the sterner *colonel*; and though he landed in Sacramento—clairvoyantly anticipating the western terminus of the transcontinental railroad—yet again, Colonel Plumbe couldn't quite give his vision life.[35]

By the 1850s his health failing and money gone, Plumbe was left with nothing but a pocket full of near misses. His great gift as well as his misfortune was to foresee a world frustratingly unreachable despite the near clarity it attained, if only to his own eyes. In his *Memorial against Mr. Asa Whitney's Railroad Scheme* (1851), written to dissuade Congress from granting exclusive rights of monopoly to Whitney for the construction and operation of a transcontinental link, Plumbe paints a wondrous picture of everything that the railroad could bring, including peace to the "whole human family"—but only if run for the benefit of the commonwealth. Realizing that the prophesy of harmony in his *Memorial* might have taken on a too-rosy artistic license, Plumbe asked his audience to see matters as he saw them, to "decide whether, with such a soil and climate as those of the Great West, I am not fully warranted in painting this picture I have presented to your view as a true *daguerreotype.*"[36]

Plumbe was right, or at least partly so, for far from unrealizable, his plans for the transcontinental railroad—including where the eastern and western termini would be, and how the federal government could subsidize private construction through the sale of alternating sections of public land—proved in time to be heartbreakingly prescient. The only thing that ultimately proved unrealistic was his belief that the railroad should benefit us all.

Unable to live in the bright future and unwilling to live in the present, Plumbe committed suicide in 1857.

Had he held on a scant five years longer, Plumbe would have seen his obsession with getting the government to support a transcontinental railroad come to fruition in the Pacific Railway Act. Then again, maybe it's Whitman who most accurately foretold what was to come: speaking of the images Plumbe had graven, the young Whitman wrote, "Plumbe's beautiful and multifarious pictures all strike you . . . with their *naturalness*, and the *life-look* of the eye—that soul of the face! In all his vast collection . . . we notice not one that has a dead eye."[37]

4

"To say the morning sun reveals untold beauties, is commonplace indeed. Words cannot express or describe it," wrote Russell in two sentences that perform their own argument—opaque literary boilerplate obscuring visual revelation.[38] It was 1870, within a year of making *1000 Mile Tree*. I read his statement as both an artistic manifesto and an interpretive challenge: words cannot express or describe whatever it was that Russell preserved with his camera; yet neither he, nor those nineteenth-century Americans who used his images, nor any of us who have since wrestled with what his photographs might mean, none of us has been able to divorce word from image.

In his own time, Russell's photos were used for a bewildering variety of purposes—to celebrate the railroad, most obviously. But they also slinked their way into explorer Ferdinand Vandeveer Hayden's *Sun Pictures of Rocky Mountain Scenery* (1870), a work that Hayden intended to be rampantly pro-expansionist, and which includes *1000 Mile Tree*, but which instead winds up coopted by a morbidly, catastrophically sublime epilogue, written by Hayden's mentor, the geologist J. S. Newberry, who detailed the inevitable cyclic extinction of human civilization despite the silly self-indulgence of technological utopians.[39] Russell's photos also appear in stereographic sets—sometimes credited to Russell and sometimes not—titled and available for individual sale. They were used as lantern slides by "Professor" Stephen J. Sedgwick (his honorific, like Plumbe's, was also self-awarded) in his popular series, *Illuminated Lectures across the Continent.* And they show up as curiosities to help advertise George W. Williams and Company's Carolina Fertilizer.[40]

What meaning does *1000 Mile Tree* hold? It depends where you look. If Russell's photos could be coopted for nearly any purpose, perhaps the photos themselves mean nothing at all, and whatever story we think they might tell is determined by their titles, their captions, their encircling texts. If so, it doesn't matter whether it was a railroad or heap of manure one used Russell's images to sell; they were used to *sell*—they are, inescapably and always, the visual fixers of Manifest Destiny, of capitalism's emerging culture industry. If one takes this approach to photo criticism, standing with the critical theorists Max Horkheimer and Theodor Adorno in arguing that "all mass culture under monopoly is identical" in its worship of wealth, then understanding any photograph paid for by the Union Pacific Railroad as a billboard for capital makes sense.[41]

Then again, perhaps individual images do have some sort of socially derived meaning, even though that meaning often finds itself massaged by different users—the photographer, his employer, the press, historians—into one interpretation or another. In that case, the best method for determining meaning, however rickety, might be to shift the angle of inquiry from the object to its use, to ask not "for whose interest was this photo produced," but "how has this photo been used to tell which stories," to then arrive at meaning by tracing its passage through a network of production, marketing, and reception, a method that locates images in a complicated and flexible visual economy. Though such a practice lacks the sure-bet definitiveness of subordinating image to word, it nonetheless opens the possibility of multiple, contested meanings and reserves for the silent image an air of mystery.[42]

In 1869, in Weber Canyon, a ray of light chased a shadow into the thin depth of an egg's white, there to stay stuck, and I, moth drawn, can't seem to help but follow, to gape at Russell's gracefully bent aesthetic conventions, mesmerized by the iconographic play evident in the photograph itself. Though reading images either for the fingerprints leftover from the ideological molding of a supposedly omnipotent culture industry, or as artifacts in a visual economy have tremendous strengths, each method ultimately shifts the privilege of historical knowledge from the image to something else — to text, or social network, or economic paradigm. But what — I want to know — about the picture, what about *1000 Mile Tree* itself? After all, there's little else to go on: in the 1870s, Russell's business partner, O. C. Smith, put a match to all his business records, destroying them in a great and sudden flash.[43]

Light chases shadow, as I tow behind, ensnared by *1000 Mile Tree*'s aura, a spectral web, "a gossamer fabric woven of space and time," as Walter Benjamin once described it, that intangible and yet deeply felt sense emanating from the photograph of an almost-living interconnection: every photo has a particular birth place and date, its own time, and yet, like people, they forever live in the present.[44] My disciplined, distanced, critical, university-learned objectivity, whose first rule is to never let the past intrude into the present, can't withstand the force of imagination, and there's no escaping the pull of the image. Images live. And it may be that the dream of standing free, outside of the frame, beyond the lure of a photograph, outside the demands of the past, in a world in which words restrain the anarchic possibility of too-real pictures, it may be that all of this is a fantasy of control.

What if we let go and embrace a different way of seeing? What if we look too long, fall in love, remain ensnared by *1000 Mile Tree*, let it drag us with its spider's silk deep into all the other images that trace Russell's life; what if we dwell among his photos and treat them as if they are alive — which is how many of us actually behave around images: the photograph of a loved one in the wallet; the picture of one's lost youth; the inspirational image of graduation or marathon finish on an office desk —

what if we ask them what they want, and listen to them as they speak to us and as the speak to other images and to the tropes of our vision? What if we focus on the photos and see how they do what they do, as individuals and as groups, according to their own offbeat logic. What if, instead of a text-based critical distance, a "critical idolatry"? Instead of plotting the circulation of things through a visual economy, what about watching how nearly living images inhabit a visual ecology?[45]

5

Though it's hard to say what effect his work had on the photographers of the later nineteenth century, Plumbe's descendants continued to bargain between death and life eternal, between present, future, and past, and to imbue their inert glass plates and mixtures of chemicals with a living spark. During the Civil War, some critics saw that the camera seemed to affect the war's outcome. "The assistance rendered the national arms during our late war by photography," wrote the editor of *Anthony's Photographic Bulletin* in 1882, "was much greater and very much more important in its results than most people imagine."[46] The photographer he had in mind was A. J. Russell, who was the only photographer actually enlisted by the US Army to shoot the Confederates with a camera. Russell would later write of his experience, "the memories of our great war come down to us and will pass on to future generations with more accuracy and more truth-telling than that of any previous struggle."[47] Characteristically, he gives us no hint as to what story those frozen memories told.

Russell was certainly a patriot. When war broke out, he painted a panorama—"the most important work of its kind in progress," trumpeted the Hornellsville *Journal*—depicting battle for the purpose of encouraging enlistment in the Union Army.[48] Yet brushes and verisimilitude were not enough for the inflamed Russell, and he also formed his own company of recruits, the 141st Regiment of New York Infantry.[49] Then, in February 1863, and for reasons that are not quite clear, Russell was personally chosen by General Herman Haupt, head of the Construction Corps of the US Army Military Railroad, to head the army's one-man Photographic Corps.[50] Whatever Russell knew of photography in 1863, it couldn't have been much, because his first act in his new commission was to hire one of Matthew Brady's assistants, the shadowy Egbert (or Edgar) Guy Faux (or Fowks, or Fowx)—no one is quite clear on the orthography of his name. Faux taught Russell the secrets of faking real life with camera and collodion for $300—the same fee a wealthy Northern gentleman paid a poorer stand-in to take his place on the front. The same fee that, in 1869, would buy a round-trip train ticket from Boston or New York to California and back.[51]

Faux, by all appearances, earned his pay, and many of Russell's Civil War pictures are clear, undistorted, fine-grained illustrations of General Haupt's prowess in bridge building, rail repairing, and Confederate infrastructure destroying, which Haupt used to help document his memoir, *Reminiscences of General Herman Haupt* (1901). The general's version of the war is, unsurprisingly, a glorious one, full of long-shot victories thanks to his own bravery and ingenuity in commanding the Construction Corps. And so it's surprising that a number of Russell's photographs deflate the overheated triumphalism of the Civil War, and even highlight the silliness of Haupt's braggadocio.

If, in the 1860s, you had wanted to make a heroic picture to stiffen the spines of those who would soon die charging into enemy bullets, you probably couldn't have done much better than to emulate Emanuel Leutze's famous 1851 painting of Washington crossing an ice-choked Delaware River. (He was the same painter whose *Westward the Course of Empire Takes Its Way* (1862) would later grace the US Capitol.) Leutze pictures a magnificently dressed general defiantly standing in a desperately overfull boat, eyes fixed on a triumphal future. It was an utter sensation upon its debut: copies were engraved and distributed widely, and so Russell was undoubtedly aware of it. But when he photographed Haupt, we see a tall man precariously crammed into his floating contraption, looking warily, and none too bravely, off to his right, paddling with a ridiculously outsized rowboat oar. He looks brittle and unsteady and scared that the jagged pylon jutting from the frame's right hand is about to scuttle his craft.

FIGURE 14. Emanuel Gottlieb Leutz, *Washington Crossing the Delaware*, 1851 (post-cleaning). Oil on canvas, 149 × 255 in. (378.5 × 647.7 cm). Gift of John Stewart Kennedy, 1897 (97.34). Image copyright © The Metropolitan Museum of Art. Image source: Art Resource, NY.

FIGURE 15. A. J. Russell, *Gen'l H. Haupt.* ca. 1862. Library of Congress, Prints & Photographs Division, LC-DIG-ppmsca-10341.

Leutze's Washington willed his way into history while Russell's Haupt paddled around its edges; but what of those who followed their leaders? On May 3, 1863, during the Battle of Chancellorsville—a continuation of the long, and very bloody December 1862 Battle of Fredericksburg— Russell took what has become one of the iconic images of the Civil War; again, it is composed to gently question the glory and triumph of war. The scene is peopled solely by corpses. In most of Russell's photographs—both during the war and out on the range with the Union Pacific—there are living humans in the frame. But here, the closest we get is a dead soldier, shot in the head, bloody face tilted skyward in death as if in his final extremity he looked for a salvation that was nowhere to be found. It recalls that panorama Russell painted at the outset of the war, which one reviewer noted, dwelled upon the "resignation of the wounded and the dying, and the marble-cold expression of the up-turned faces of the dead," except that, unlike the panoptic panorama, the camera's eye is ours, and our presence is effected in the photo.[52] We, the audience, stand in the scene, the sole surviving witnesses behind the wall.

It's also a landscape image, and the surrounding trees and stonewall are crucial parts of the photo. On any other day, Marye's Heights would be the scene for a picturesque view of Fredericksburg; but on this day, preserved forever, a horrible stillness allows one to question the bloody orgy that occurred only moments before Russell uncapped his lens, which is particularly suggestive given that that his photo captured the corpse of a Confederate. During the first battle of Fredericksburg, in December 1862, the Union had suffered terribly, and its losses far outnumbered Southern casualties; Marye's Heights was the site of some of the most desperate fighting. The North lost that battle, but in May 1863, when the battalions of blue-coated soldiers returned, the Union momentarily overran the Heights. Russell was immediately on the scene, making his view, and he was later praised for the "eye of an artist in choosing his stand points for the most favorable presentation of the objects or scenes taken."[53] Again, the North ended up losing Chancellorsville, which makes Russell's photo all the more remarkable: rather than demonizing the dead as devilish secessionists who got their fair

FIGURE 16. A. J. Russell, *Stone wall, rear of Fredericksburg with rebel dead, May 3d, 1863*, from *Military Installations, Activities, and Views, Washington, D.C., Richmond, Va., and Vicinity*. Library of Congress, Prints & Photographs Division, LC-DIG-ppmsca-11747.

due, Russell, by placing us on the Confederate side of the stone wall, asks us to identify with the slain, with *them*, not in their political aims but in the snuffing destruction that war brought to living bodies on both sides of the line, over ground dappled gray and blue and red, while the silent stone walls and trees, once the markers of picturesque harmony, but here figured as failed protection from harm, stand as tombstones and grievers at a deserted funeral.

Even as events tilted decisively toward Union victory, Russell continued to make pictures suffused with ambivalence. One of his starker images is *Engine "Government" down the "Banks," near Brandy, April 1864*, and at first glance it seems like the sort of conventional photograph that one would expect from a man whose commission was to make images of war and trains. But the title is a dead giveaway, and the photo literally tells us that government—the ship of state—had foundered. The basic composition of the photo is exactly that of *Stone wall, rear of Fredericksburg:* two parallel diagonal lines—wall and train, trench and tracks—guide the eye, though in *Engine "Government"* the people, though they seem powerless to help, and those off to the left squat on their heels in the dust, waiting for an unknown something, at least the people live. As in *1000 Mile Tree,* the figures are small, generic; their size and lack of specificity are ridiculously outmatched by the sheer heft and intricacy of the overturned train, whose guts—the most visually complex part of the photo—lie uncovered and exposed for postmortem examination. Our eye is meant to follow the tracks—Russell has placed his camera centered over the right rail—into the empty distance, only to be interrupted by the human knot in the background. There is no sense that the road disappears into the horizon, often a sign of the future's promise; the very landscape itself seems foreboding. Its barrenness and uniformity, bridged only by a technological wonder at that time far beyond help, threatens either to swallow the characters in the photo or to condemn them to wandering in a featureless desert.

Looking back, in 1882, on his role in the war, Russell described how

FIGURE 17. A. J. Russell, *Engine "Government" down the "Banks" near Brandy, April, 1864,* from *Military Installations, Activities, and Views, Washington, D.C., Richmond, Va., and Vicinity.* Library of Congress, Prints & Photographs Division, LC-DIG-ppmsca-08273.

his photographer friend Thomas Roche was suddenly bombarded when making pictures of the battle at Dutch Gap Canal: "Down... with the roar of a whirlwind [came] a ten inch shell, which exploded, throwing the dirt in all directions." Miraculously unscathed, Roche went on to make a number of pictures of the scene, to fix the lightning-quick moment of death in egg-white permanence. Russell paints a complicated, enraptured verbal picture that emphasizes earthquakes and ordinance tearing at the ground, as well as an almost organic liveliness: "Shells flying in all directions, leaving their trails of fire and fading away only to be replaced by others.... The whole world seemed alive; every road was teeming and the call to arms seemed to find a response from every foot of the ground"[54] Accompanying his verbal description is the quick pencil sketch in figure 18. In some ways it's a sublimely nationalistic image of heroism: a lone photographer struggling at his trade, unarmed except for his camera and glass plates, the smoke and dirt of

FIGURE 18. Russell's sketch of Dutch Gap Canal. A. J. Russell, "Photographic Reminiscences of the Late War," 1882.

an exploded shell obscuring his vision, enemy troops in the distance hard at their killing. But I think what captures the eye, and what Russell must have been at pains to include, are the tombstones that occupy the left foreground of the picture. They're crooked, jagged, thrown up hurriedly. If Russell's prose tells us that the photographer cheated death, his drawing seems to be more about striking a Faustian deal with mortality: a photographer's life spared in exchange for death immortalized. It's an obsession and a concession that Russell seems to have been forever trying to work out of his system.

Yet the starkest of Russell's wartime photos come not from the field of battle but from the places that were once thriving, vital city centers, places like Richmond, whose destruction Russell documented compulsively, forcing us to consider that thing so unnaturally haunting: a city bled of its people. There are any number of ways that he could have chosen to take these photos. He could have opted for triumph, with a Union soldier waving the US flag atop a captured arms depot. Or hope, with gray and blue coming together to rebuild. Or nationalism, with commerce humming along despite the war. Instead he chose empty streets, occluded by rubble; gutted buildings whose misanthropic hostility is emphasized by the careful placement of two lone individuals standing beneath an isolated, ironic lamppost (upon what will it cast its feeble light?); a river curving gently between factories and under the graceful arch of a stone bridge that a decade before would have evoked pastoral associations of harmony—but this isn't a peaceful picture, as the weeping willow, long understood as death's sylvan stand-in, occupying the center of the frame would suggest. The streets are empty, the jagged ruins of a mill shiver skyward, and harmony seems to have been buried when the roof caved in.

6

In the best photographs, there's a shock of the new. They confront us in some way; they force us to look. Or they simply burrow in, enticing us with the hope of revelation to return and return again. Some photos

FIGURE 19. Ruins in Richmond, Virginia. A. J. Russell, *Military Installations, Activities, and Views, Washington, D.C., Richmond, Va., and Vicinity*. Library of Congress, Prints & Photographs Division, LC-DIG-ppmsca-08232.

FIGURE 20. Ruins in Richmond, Virginia. A. J. Russell, "Military Installations, Activities, and Views, Washington, D.C., Richmond, Va., and Vicinity." Library of Congress, Prints & Photographs Division, LC-DIG-ppmsca-08229.

are pleasurable, others cut; some are inviting, others repulse; some are calm, others violent. But every good photo asks something of its viewers: images, too, have desires, if only to be gazed at, and when we listen to them, we do so on ground that is partly ours and partly theirs, a space in between subject and object, real and fantastic, fact and meaning, prose and poetry, a kind of no-man's-land common space where dialogue between viewer and image becomes a real possibility. When we look closely, we do so because we expect something: we expect communion. This is the space of revelation.[55]

One of the things that makes Russell's photographs so enchanting is that he wasn't blindly taking snapshots of whatever struck his fancy—he had to work for them: loading hundreds of pounds of gear onto the backs of strong-willed mules, or carrying gallons of pure water, in one case for seventy miles, not for drinking, but to slake the delicate thirst of his images, for whom the alkaline desert water would spell death.[56] And then, to make a single photo required, at the bare minimum, half an hour of unpacking equipment, preparing the glass, exposing film, and developing his image.[57] There was nothing snap about his shots, and you can see, 140 years later, deliberation thrumming its way through his images. Yet their meanings, as always, are never apparent. We have to work for them, and one way to begin appreciating Russell's visions is to restore his images to their native ecosystem: as one among the first wave of western landscape photographers, Russell was on the very cutting edge of artistic photographic aesthetics, borrowing from the dominant visual landscape conventions of his time—the sublime, the picturesque, and the beautiful—and then shaping them to fit his own vision.[58]

A history of aesthetics can help us appreciate Russell's work, though we must take care to not cage his images in with precise definition. Besides, the boundaries setting off one aesthetic from another have always been fluid. In the United States, the picturesque historically absorbed both the sublime and the beautiful, and so each convention quickly shades into the others, often in the very same image.[59] Artists themselves have long blurred aesthetic distinctions, reforming each to confound, tease, and delight an audience's expectations.[60] In other words, aesthetics have a history—they're culturally unstable, they change over time; they evolve. Yet, loose distinctions between aesthetic categories can be helpful, because such territorial markings gesture to different, implicit bodies of ethics, to what is valuable about the world, what and who belongs in it, which actions are virtuous and which debauched.[61]

And so the sublime is typically an aesthetic of power, domination, separation, the irrational, faith.[62] It was originally applied to literature and biblical history, but, owing to the two great earth-moving scientific discoveries in the seventeenth century—the twin discoveries that the cosmos are in a state of constant change and that the earth is almost unknowably old—the sublime came to inhere most strongly in the land.[63] To put it another way, Earth itself came to take on the mysterious, infinite omnipotence traditionally reserved for the Father, Son, and Holy Ghost. Edmund Burke, the Anglo-Irish statesman and philosopher most often associated with the sublime, thrilled to things that challenged his very survival, and in the mid-eighteenth century, Burke began to theorize that a brush with annihilation provided the titillating impact of sublimity. "Whatever is fitted in any sort to excite the ideas of pain, and danger," he wrote, "whatever is in any sort terrible . . . is a source of the *sublime*."[64] This is the twist: Burke's sublime depends not on human might but on our collective powerlessness in the face of an almighty nature. If there's a silencing, it's of the human voice; if there's domination, it's of nature's infinite complexity overwhelming simple, blind, human dust motes. For Burke, the sublime lived in the enchanted physical landscape but was always capable of forcing its way into the puny world of human affairs. With a flick of its wrist, nature could wring all kinds of exquisite agony from its delicate object: general darkness and obscurity, manifestations of great power, the threat of privation, vastness, infinitude, great difficulty, magnificence, great light, loud or unfamiliarly terrible noises, bitter smells and awful stenches, pain— these were the landscape characteristics most calculated to cause the sort of astonishment that could shock from a person his sense of security.[65] And yet there's also a quieter sublime that often accompanies the powerful Burkean one, the sublime of infinitude: it's the sublime of geology, of a tree that is a thousand years old, of the vast and undifferentiated ocean, of the grass-sea plains. It, too, humbles: the three score and ten years that the Bible allots to a human life is barely more than a grain of sand in the sedimented layers of deep geologic time.[66]

If the sublime is about domination, the picturesque is concerned with harmony. Most eloquently theorized in the late eighteenth-century by Englishman William Gilpin, picturesqueness rests on the reconciliation of the seeming dissonance between pure nature and human culture.[67] Neither nature nor culture is solely material, according to the picturesque, but the phenomenal expression of a higher ideal; picturesque landscapes expressed an Edenic reconciliation between humans and the natural world. Picturesque views were those that were pleasing to the eye, rather than thrilling to the emotions, and offered a vision of perfection. Uvedale Price, a vocal participant in the English aesthetics debates of the late eighteenth and early nineteenth centuries, framed the picturesque this way: though it was a visual aesthetic, closely related to painting, in the end, human-made images were simply not enough. Neither were actual physical landscapes. Instead, for Price, whose notion of picturesqueness was bound up with a belief that nature and society ought to mutually improve one another, the great power of picturesque images is that they were forever incomplete and in need of careful cultivation. Always a bit of an abstraction, the picturesque was not so much an image of how things were, as much as what they could be—the path back to the Garden. Form, lightness, balance—these are the crucial picturesque factors, and without all three, a landscape could be interesting, or even beautiful, but never picturesque.[68]

Finally, the beautiful was anything that had "that quality or those qualities . . . which . . . cause love, or some passion similar to it."[69] Picturesqueness could reside in ugly things, as long as they fit into the landscape: ruins, graveyards, or shipwrecks could all work if they blended well. But beauty was reserved for those things that exhibited pattern, regularity, symmetry, restraint, proportion—things that followed the rules, for there were to be no surprises with the beautiful, nothing awful or grand, no bitter notes that gradually resolved into sweet.⁷⁰ There is a one-on-one intimacy here, a vision of a tree that you can hug, as well as a sort of morality: hugging beautiful things is a virtuous act. As Burke put it, beauty created a sympathetic "great society" between humans and the natural, beautiful world.⁷¹

Of course, Burke, Gilpin, Price, and the artists who modified their landscape aesthetics, like Russell, are also easy marks for today's critical sledge: each has his chauvinistic moments of paternalistic, culturally elite, white, male, and Western bias, and their collective theories can be tallied up as instances of an overidealized, simplified, ahistorical, modern flight from reality that greenwashes the social and environmental impacts of human industry. The eighteenth-century theorists and their later followers often have that infuriating old-time Enlightenment confidence that they've discovered bedrock principles, that cocksureness bordering dangerously on hubris, all of which needs to be questioned. No aesthetic is simple, no aesthetic means exactly what it says, and one can find dozens of picturesque, beautiful, or sublime views used to disguise environmental degradation and social inequity. Take Thomas Cole, who, enraptured by beauty, wrote, "There is in the human mind an almost inseparable connexion between the beautiful and the good."⁷² We can ask, whose idea of beauty? Whose idea of the good? Certainly not all humans held the same judgments as Cole—why should he speak for everyone? The picturesque, too, can be quickly taken apart as an aesthetic that naturalized all sorts of destructive technology—from steam engines to sawmills—into something softer and more palatable. Even the sublime has paradoxically been used to value both the Grand Canyon and the seemingly geological dams that constrict its life-flowing river, the concussive wave of a waterfall and the book of Revelations–scale nuclear mushroom cloud.⁷³ Ansel Adams neatly highlights the problem of such aesthetics: he's a master of the sublime, but there's a triumphal nationalism to his famous landscape photos—only in America does there exist a Yosemite—a sort of self-satisfied sense of well-being that is accepting of anyone who professes to love the wild. An Adams photograph hung in the expensively decorated corporate meeting room of a multinational oil company doesn't necessarily screech cognitive dissonance, because Adams's own use of the sublime was never much of a critique of exploitation in the first place.[74] That multinational oil company could very well be a corporate sponsor of the Sierra Club or Nature Conservancy, and financial support for preservation is really all that Adam's Yosemite photos ask. So we should be skeptical of pretty pictures.

But what suspicion risks losing is the deep though imperfect humility that aesthetics like the sublime, the picturesque, and the beautiful have often helped artists struggle toward. Chalking each up to a Western culture always-already corrupted by capitalism and empire threatens entire constellations of critique, dissent, negotiation—and not incidentally beauty. That's too high a price to pay.[75]

If it's true that any aesthetic can be harnessed for the purposes of profit and power, then it's also worth noting that aesthetics are never wholly enslaved: no one unilaterally controls how an aesthetic is used, who or what it can show, how it privileges one view over another, what meaning it makes of the world—a fundamentally ungovernable nature is one of the things that makes good art always mysterious and ever vital.[76]

What's more, there's a raw current of resistance running through many formulations of different landscape aesthetics—it's what electrifies Russell's work.[77] Indeed, the very notion of the sublime, of an infinitely powerful nature, mocks that phrase conceived in arrogance, the control of nature, and has the potential to unseat the simplistic notion that all land is destined solely for human use, whether as private property or natural resource. If one agrees with Burke's conception, then the sublime, because it inheres in the natural world, can be a way of letting nature in, a way of valuing land not for its economic potential, but for its ability to make us feel, for its ability to make us human even as it denies our self-importance.[78] "I cannot describe," wrote Russell, from the mountains of Utah, "the feelings of a person when first stepping from the green carpet of grass nourished by midsummer sun upon this great winter born snow bank. At first it does not seem a reality but when you step out upon it, feel its crisp beneath the feet, and realize the change of atmosphere, it is delightful and exhilarating in the extreme. . . . From the snow bank I can see nearly one hundred miles to the north-east and west, but to the south, south-east, and south-west vast ranges of mountains shut in everything beyond. I do not know which to admire the most."[79] Movement from a landscape of growth and nourishment to one of hardship and coldness, from life to living death, having his vision at once let out and hemmed in, all of this reanimated Russell. He was reborn through helplessness. And so the sublime could bring with it a radical revaluation of places normally deemed wastelands: crags, mountains, chasms, great winter-born snowbanks—howling wildernesses standing in the way of human progress, fit only for the devil.

Perhaps because of its association with death, we could see the sublime as potentially the archetypical aesthetic of limits. But harmony, too, came with its own set of implied ethics that look a lot like what we might now call sustainability. When Gilpin promoted the "harmony of parts" in his picturesque, he was playing on the reconciling of opposites, of nature and culture: harmony means concord after all, and one can almost feel him wince when dissonant strains invaded the landscape.[80] There's a passage at the very end of the first volume of *Remarks on Forest Scenery*, written as a sort of history of the world from the point of view of an Old Testament God angered at how those created in his likeness have fouled their only nest, remarkable for its full-throated condemnation of early capitalism:

> Though man had deserted the forest as a dwelling, and had left it to be inhabited by beasts, it soon appeared that he had no intention of giving up his right of dominion over it. In a course of ages, as population increased, he began to find it in his way. In one part, it occupied grounds fit for his plow; in another, for the pasturage of his domestic cattle; and in some parts, it afforded shelter for his enemies. He soon shewed the beasts they were only tenants at will. He began, amain, to lay about him with his axe. The forest groaned, and receded form its ancient bounds. It is amazing what ravages he made in his original habitation through every quarter of the globe. The fable was realized: man begged of the forest a handle to his hatchet; and, when he had obtained the boon, he used it in felling the whole.

Gilpin, the man whose *Remarks on Forest Scenery* is an ode to beautiful trees, cannot hide his wrath at the historical destruction dealt to the world's picturesque woodlands: "In very early days, this devastation began. When Joshua divided the Land of Promise among the Israelites," and later, "The mighty forest of Lebanon, which once found employment for eighty thousand hewers, is now dwindled to a dozen trees." Gilpin's angry eye roves across the world: "The woods which covered the island of Delos had entirely disappeared.... In all the new peopled parts of America ... it is astonishing what devastation the woods of these countries have suffered.... In the West Indies.... In Barbadoes.... In the East Indies ... wherever settlement has been made, the woods have been cut down."[81]

Finally, the beautiful could foster the deeply ethical, democratic imperatives of enjoyment. "The spirit of our age," Cole wrote in his *Essay on American Scenery* (1836), "is to contrive but not to enjoy—toiling to produce more toil—accumulating in order to aggrandize." Instead of leading lives of quiet desperation, Cole argued, we need to open our eyes to the beauty that surrounds: "He who looks on nature with a 'loving eye,' cannot move from his dwelling without the salutation of beauty.... The delight such a man experiences is not merely sensual, or selfish ... he feels a calm religious tone steal through his mind, and when he has turned to mingle with his fellow men, the chords which have been struck in that sweet communion cease not to vibrate."[82]

7

War allowed Russell to hone his photographic craft; it also proved to be the event that finally made Plumbe's dream of a transcontinental railroad come true. Congress passed the Pacific Railway Act in 1862 (and again in 1864, with even with even larger handouts in cash and land to the Union and Central Pacific Railroads), and by the time Robert E. Lee surrendered to Ulysses S. Grant at Appomattox, the railroads' gangs had begun work, ready to apply wartime construction lessons to annihilating the distance separating the Mississippi from the Pacific Ocean. The end of the war found Russell back in New York, painting landscapes and panoramas of, among other things, explorations of the polar seas and Dr. Livingston's travels in Africa.[83] But soon thereafter—thanks, most likely, to General Haupt's engineering—the Union Pacific hired Russell as its official photographer.[84] Loading up his photographic gear and heading to Wyoming in the summer of 1868, Russell began making hundreds of photographs along the Union Pacific's route, both in ten-by-thirteens and in smaller, side-by-side stereo-view formats. Though the cheaper stereo views enjoyed the widest distribution, his large-format photographs—with their superior quality and uncanny ability to mirror the world—enchant. It was a selection of fifty of these large images that appeared in *The Great West Illustrated* (1869), published by the Union Pacific Railroad a month before the laying of the last rail at Promontory Point, which constitutes what Russell's recent biographer, Glenn Willumson, has called "the most famous visual production of the transcontinental railroad and one of the most important photographic representations of the American West in the nineteenth century."[85]

Before cracking the first page of *The Great West Illustrated* and discovering the golden plates within, one can tell by its cover that it's a testament to the nearly miraculous feat of spanning a continent. The thing weighs fifteen pounds, and every aspect of the album, which cost somewhere between $50 and $75 nineteenth-century dollars—or between roughly $800 and $1,200 today—breathes of assured luxury.[86] "Published by Authority of the Union Pacific Railroad Company," intones the opening page, before the preface continues: "Few persons, and only a small number of Scientific Expeditions, have traversed the Great Central Belt, and its History, Geography, and Geology are almost unknown. It is therefore believed that the information contained in this Volume, and which will be continued in those that follow, is calculated to interest all classes of people, and to excite the admiration of all reflecting minds as the colossal grandeur of the Agricultural, Mineral, and Commercial resources of the West are brought to view."[87] The language testifies to a self-satisfied celebration of capital and its glittering future, ensured by the "colossal grandeur" of natural resources opened by the

FIGURE 21. A. J. Russell, *Malloy's Cut*, 1869. From *The Great West Illustrated in a Series of Photographic Views Across the Continent* (New York: Union Pacific Railroad Co., 1869). Yale Collection of Western Americana, Beinecke Rare Book and Manuscript Library.

railroad. Even the decorative sketches that grace each page—pleasant little locomotives that descend from the extreme right and left, like the chubby cherubim that inhabit classical allegorical painting—seem to set the stage for a corpus of photographs that we expect to be rousing.

Instead, within the first pages, we get images such as *Malloy's Cut*, a stuck-tight view jammed between parallel walls of a cut whose gritty geological texture contrasts oddly with the scene's otherwise blandness. *Malloy's Cut* hems our vision in, directs it toward the background where the rails disappear around the far-right corner of the embankment. Besides the lone, distant knob centered between the gun sights of the cut, there isn't much, save the solitary figure, for our eye to rest on; and he, unknowable, turns from us in a three-quarters pose, the classic position meant to suggest veiled mystery.[88] Who is he? How did he get to this place? How will he leave? What is it that he looks at? The accompanying text is at a loss to provide any narrative clue beyond the purely extraneous and seems vastly deflated from the bombastic prose of the preface: "The road here is cut through a mass of disintegrated granite. Malloy's Cut is situated two miles east of Sherman's Station, the summit of the Union Pacific Railroad." Even this literary shard feels disconnected from the image: it appears in the annotated table of contents, separated from *Malloy's Cut* by nine massive pages. One gazes at Russell's view alone and unaided.[89]

A few pages farther on, further west, we arrive at *On the Mountains of Green River*, as acrophobic as *Malloy's Cut* is immuring. Russell has posed three men and three women on the edge of a butte while the camera seems to hang in midair, giving the voyeuristic audience a paradoxically unreliable God's-eye view of whatever drama is in the midst of unfolding. The text again seems woefully out of place, almost as if meant for another image: "The standpoint for this view is nearly two

FIGURE 22. A. J. Russell, *On the Mountains of Green River*, 1869. From *The Great West Illustrated in a Series of Photographic Views Across the Continent* (New York: The Union Pacific Railroad Company, 1869). Yale Collection of Western Americana, Beinecke Rare Book and Manuscript Library.

thousand feet above the Railroad, which can be seen, winding through the bottom lands, three miles away; farther off can be seen the dim outline of Green River City. This town is built of unburnt brick, and when this view was taken contained two thousand inhabitants." If the caption hadn't called these details of human settlement to our attention they would be all but invisible, swallowed up by distance and collodion. In both *Malloy's Cut* and *On the Mountains of Green River*, Russell framed his images so that natural features trump human agency: there's nowhere for the lone figure standing beside the tracks in *Malloy's Cut* to go, no place to find asylum. The three couples in *On the Mountains of Green River* are poised on the brink of a dead world's edge, far beyond both the life-sustaining river and the railroad—the putative star of the show—with no way to descend. And in both, a deep geological sublime dwarfs human impermanence—the invisible town of unburnt brick, the riderless rails.

An evanescent, if living, human shadow cast briefly across the strata of catastrophic geology is the main character of one of Russell's more famous photographs, *Hanging Rock, Foot of Echo Canon*, another image that appeared in *The Great West Illustrated*. Again, the annotated table of contents seeks to guide the viewer's eye: "This mass of conglomerate rock, overhanging its base nearly fifty feet, and forms the foundation of a bluff nearly one thousand feet in height. It Overlooks Echo City. The Union Pacific Railroad winds around the bluff's base. From its top can be viewed some of the grandest scenery on the road."[90] Again, we get the rubble pile of distracting verbal description absent of even a scrap of meaning; for that, we must look at the photograph itself, and what is immediately striking about *Hanging Rock* is the great detail of the geological features compared to the built environment. We can see every pebble that makes up the conglomerate's hulk, whereas the clapboard house, barn, and outbuildings in the background seem generic and fleeting. Again, Russell has posed a lone figure in the scene, and again he turns away from our enquiring eye. *Hanging Rock* vibrates with visual tension, and if one gives oneself over to the photo's logic, the megalith is only moments away from crashing down—unable to support its own

FIGURE 23. A. J. Russell, *Hanging Rock, Foot of Echo Canon*, 1869. From *The Great West Illustrated in a Series of Photographic Views across the Continent* (New York: The Union Pacific Railroad Company, 1869). Yale Collection of Western Americana, Beinecke Rare Book and Manuscript Library.

unbalanced weight—not only upon the hapless sitter urged to go west and grow up with the country, but also upon the small Jeffersonian settlement as well, snuffing out the dream of western expansion.[91]

Out here, in the American West of 1868, Russell resumed his wartime work of developing compositional techniques that undermined the longed-for triumphal tales of progress, and one of the ways he did this was by refashioning landscape aesthetics. Though he doesn't seem interested in the beautiful—perhaps it, too, was killed off at Fredericksburg— he continued to develop his takes on the picturesque and sublime in a bath of irony, yielding a forlorn antihuman natural sublime, in which Burke's thrilling annihilation is replaced by the futility of Progress, as well as a problem picturesque that denies the possibility of harmony (a paradox!), aesthetics that militate against the possibility of humanity standing in mastery over a tamed nature.[92] Dozens of Russell's western landscapes show a lone, anonymous person seemingly stranded in a silent wilderness; or, as in the case of *1000 Mile Tree*, if there are groups, they are almost always abstract stand-ins, rendered insignificant by their distance from the camera's lens. Railroad technology fares no better: it's rare to see an image of a locomotive, the iconic embodiment of rushing iron and roaring steam—of modernity.[93] Instead, Russell was at pains to capture a landscape that swallowed up its inhabitants in infinite space and drowned them in deep time, and his images either register a minimal claustrophobic depth by compressing the space of the image into a flat plane, as in *1000 Mile Tree*; or they block the free passage of the audience's eye as in *Engine "Government" Down the "Banks"* and *Stone wall, rear of Fredericksburg*; or, as with *Malloy's Cut*, they let the audience entirely out into a vast nothing, free to experience the full weight of space chaotic, silent, and still.[94] Together, these techniques give rise to a landscape sensibility that prizes stasis over motion, stillness over noise, melancholic indirection over purposeful action, isolation over society—a sensibility that Leo Marx has called a "'worst case' of the harsh, unadorned truth" about empire: the star of Progress set in the West.[95]

Indeed, Russell's views seem jarring and strange when viewed alongside the more famous advertisements for Manifest Destiny. By the late 1860s, there was a tremendous iconographic bank of celebratory train-in-the-landscape images that Russell could have drawn from, including Andrew Melrose's sublime *Westward the Star of Empire Takes Its Way—Near Council Bluffs, Iowa* (1865), which locates itself at the Union Pacific's start in Council Bluffs. Here the train's star headlight cruelly screams down upon us, its glare obscuring the great force just behind, hinted at by a lick of red fire shooting from the locomotive's smoke stack to taste the night sky. Space and time, and the deer that stand in the way, even the audience, are in the very midst of being annihilated— it's fatal to stand in the way of progress, the painting seems to say. But for those who follow behind, or step to the side, there's the promise of a successful homestead, complete with three cows, wash hanging on the line, a family, and land. *Westward the Star of Empire Takes Its Way* is at once an aggressively sublime image of the unstoppable engine of

FIGURE 24. Andrew Melrose, *Westward the Star of Empire Takes Its Way—Near Council Bluffs, Iowa, 1865*. Autry Museum, Los Angeles, 92.147.1.

history—the Union Pacific, Manifest Destiny, the United States—and the modest agricultural wealth that many hoped would follow in the train's wake.[96]

Then, in 1867, Russell's fellow Civil War and railroad photographer, Alexander Gardner, made a photo also entitled *Westward the Course of Empire, Laying Tracks 300 Miles West of the Missouri River* as part of a campaign to convince the federal government to fund a second, southern transcontinental railroad route, the one preferred by the Union Pacific Railway, Eastern Division, competitor to Russell's Union Pacific Railroad. Though it's not as blatantly idealized as Melrose's work, it nonetheless prominently features an engine on its way to California, accompanied by a gang of men who are busy grading and laying track. There's a singleness of purpose in the image, one that unites machine and human, and perhaps even the landscape: seemingly by divine fiat there is not a mountain to be bored through or a valley to be filled anywhere on the horizon.

The picturesque, too, could be bent to serve Progress, as Jasper Francis Cropsey's *Starucca Viaduct, Pennsylvania* (1865) shows. There's a deep space here, and the train is almost invisible in the distance—if not for the elongated white plume of smoke, which resonates with the

FIGURE 25. Alexander Gardner, *Westward the Course of Empire Takes Its Way: Laying Track 600 Miles West of St. Louis, Missouri, October 19, 1867*. Albumen silver print, 13¹⁄₁₆ × 18¾ in. (33.2 × 47.6 cm). The J. Paul Getty Museum, Los Angeles. Digital image courtesy of the Getty's Open Content Program.

peaceful white clouds, sleepy village smoke, and the smoke from the campfire in the foreground, the train would be all but lost. The train tracks embrace the town, and the wonders of modern engineering—the bridge crossing the river, the train, the viaduct—appear every bit as natural as the maples, pines, and birches gracing the foreground.

A deep space that invites the viewer in, or alternatively, a space aggressively annihilated by a sublime locomotive; an ever-present engine, or, at the very least, its smoking trace; people happily working together on the frontier; trains and infrastructure that harmonize with the natural landscape and their human viewers; a sense of noise, possibility, and above all, purposeful movement, always movement: these are the conventions that Russell should have drawn upon if he was furthering the cause of capitalism and expansion.[97]

To be sure, not all the photographs in *The Great West* unsettle. There's the Mormon Tabernacle in Salt Lake City; the sublimely named Death's Rock and Devil's Gate; and a few classically picturesque shots of men

FIGURE 26. Jasper Francis Cropsey (American, 1823–1900), *Starrucca Viaduct, Pennsylvania*, 1865. Oil on canvas, 22⅜ × 36⅜ in. (56.8 × 92.4 cm). Toledo Museum of Art. Purchased with funds from the Florence Scott Libbey Bequest in Memory of her Father, Maurice A. Scott, 1947.58.

fishing in a gentle river. And not all of the photographs are so devoid of industry or habitation. In figure 27, *Coal Beds of Bear River*, we get the first glimpse of the mineral wealth promised us by *The Great West Illustrated*'s preface. But it's a haunted picture: we see four men, arrayed around the gaping mouth of the tunnel, and if one looks closely into the flattened space of the photograph, there's a fifth—a spectral image of a headless miner just coming into the camera's vision from the depths of the tunnel. The poverty of technology can explain away what we see (without a specialized filter, one cannot get perfect exposure of both deep shadow and bright light, so one must prioritize one or the other), but nonetheless, taken on its own, the photo literally shows us a decapitated miner's ghost as the cost of technological development.[98] From ghosts in figure 27, we move to what could be, in figure 28, *Coalville, Weber Valley*, a soon-to-be ghost town, one in which a father and his two children—a fractured nuclear family—are foregrounded before the settlement of Coalville. But they're indistinct, and it's easy to overlook them, and, anyway, the main attraction seems to be the wide road through the puny settlement, a road completely empty and that goes nowhere, cut off on the left by the frame and on the right by the intercepting hills. The picture seems to attest to a culture that values movement and transience over rootedness, but is nevertheless stuck. The title, *Coalville*, clearly hearkens to the earlier photograph, *Coal Beds of Bear River*, and even visually, the town's architecture recalls the haunted coal mine: the town's inhabitants live underground in sod-roofed houses and dugout cabins.

And so those original opening words from *The Great West Illustrated*—that Russell's sun pictures are "calculated . . . to excite the admiration

FIGURE 27. A.J. Russell, *Coal Beds of Bear River*, 1869. From *The Great West Illustrated in a Series of Photographic Views across the Continent* (New York: Union Pacific Railroad Co., 1869). Yale Collection of Western Americana, Beinecke Rare Book and Manuscript Library.

FIGURE 28. A. J. Russell, *Coalville, Weber Valley*, 1869. From *The Great West Illustrated in a Series of Photographic Views across the Continent* (New York: Union Pacific Railroad Co., 1869). Yale Collection of Western Americana, Beinecke Rare Book and Manuscript Library.

of all reflecting minds as the colossal grandeur of the Agricultural, Mineral, and Commercial resources of the West come into view" — are wildly off the mark, almost as if they describe some other book filled with Currier and Ives lithographs. It seems that the images were nearly afterthoughts in the mind of whomever put the book together: different copies of *The Great West Illustrated* contain different views, and it's the literary text that carries the burden of telling the tale of western riches that the Union Pacific so wanted to spread. Not the photos. The Union Pacific originally planned to issue two volumes of *The Great West Illustrated*, yet it stopped after one; Willumson has discovered that the corporation seems to have parted ways with Russell by 1870, though the reasons remain mysterious.[99]

I'm inclined to believe that the photographic evidence of agriculture's absence, and an aesthetic that envisioned settlements vanishing in the vast distance, lying fallow and unpopulated, and in danger of

FIGURE 29. Currier & Ives, James Merritt Ives, and F. Palmer, *Across the continent, "Westward the course of empire takes its way"* (New York: Currier & Ives, 1868). Library of Congress, Prints & Photographs Division, LC-DIG-ppmsca-03213.

being crushed by an inhospitable environment, was too much for the corporation.¹⁰⁰

If whatever Russell's photos mean refuse to be corralled by the text of *The Great West Illustrated*, if these images seem to enact desires of their own and gesture beyond the comfort of the book's covers, even beyond their own frames, to other images, other conventions, a wider visual ecology—if his images break away from the text—then we should follow them.¹⁰¹ Though *1000 Mile Tree* doesn't appear in *The Great West Illustrated*, it was nonetheless conceived of with the rest of Russell's western views, lived among them, and marked by the Russell family resemblance.

8

Irony, the antihuman sublime, a problem picturesque—to these defining characteristics of Russell's landscape aesthetics I would add one more: an obsession with the iconography of sacrifice. It's there in *Dial Rock, Red Buttes, Laramie Plains*, in which another lithic column stands like a sundial, marking out time in the center of the picture. On its right side is a small figure who, unsurprisingly, turns his back to us. He's watching something, tautly, that we can't see: he's crouched behind the shoulder of the rock for protection or concealment; he's got his rifle a muscle twitch away from being shouldered, aimed, and fired. The right side of the frame is where all the action is, and so one might miss what is on the left: a human ribcage, backbone, and pelvis nestled among the stones, a formerly living person for whom Dial Rock was not protection, but tombstone.[102] *Dial Rock* is one of the tensest photographs in all of Russell's landscape oeuvre, and in 1868, at the end of a decade that saw not only the carnage of Civil War but also the Wiyot Massacre (1860), the Chivington Massacre (1864), the Washita Massacre (1868),

FIGURE 30. A. J. Russell, *Dial Rock, Red Buttes, Laramie Plains*. From F. V. Hayden, *Sun pictures of Rocky Mountain Scenery* (New York: Julius Bien, 1870). Library of Congress, Prints & Photographs Division, LC-USZC4-11377.

FIGURE 31. A. J. Russell, *Dial Rock, Red Buttes, Laramie Plains*, detail.

and at least eighteen other massacres of American Indians by the United States, this image of unresolved violence resonated. Russell pauses the drama a gunshot before its denouement: Will the remaining living man carry the day, or will he wind up another pile of bones somewhere out on the lone prairie? Will the project of Manifest Destiny, for which the gunslinger seems to be agent, metonym, and sacrifice all at the same time, triumph, or will its time, too, run out at the foot of Dial Rock?

Sacrifice—it's a word rarely spoken by the cheerleaders of Progress. And yet, Russell was hardly alone in his anxiety. Indeed, in 1843 Nathaniel Hawthorne's "The Celestial Railroad" appeared in the magazine *Democratic Review*—the very same magazine in which, two years later, John L. O'Sullivan, all in a froth over the prospect of adding Texas to the United States, would popularize the term *Manifest Destiny*.[103] "The Celestial Railroad" is an acid take on John Bunyan's seventeenth-century *Pilgrim's Progress*, a text long familiar to many Americans that recounts a Christian penitent's path to salvation; but in Hawthorne's retelling, modern-day Americans, all in a hurry for the promised land, board the Celestial Railroad, a train driven by an engine that looked "like a sort of mechanical demon that would hurry us to the infernal regions" and which, as it hurtles along its iron-shod path, spews smoke on the traditional Bunyanesque pilgrims who plod along the train's right of way on their way to heaven. Meanwhile, the demonic director of the Celestial Railroad Corporation, named "Mr. Smooth-It-Away," drops his self-congratulatory passengers off at the "final station house," where all transfer to a steam paddle wheeler for the final crossing of the river Styx, which, it turns out, is actually the final step to hell. At the tale's end, Mr. Smooth-It-Away, with "a cachinnation of smoke-wreath" issuing from his mouth and nostrils, "a twinkle of lurid flame" darting out of either eye, stands on the near shore waving his passengers an ironic adieu, an efficient employee whose CEO was the devil himself.[104]

Nor was Hawthorne alone, for some of the mid-nineteenth century's keenest countermodern critics pointed to the machinery linking annihilation to sacrifice. "We do not ride on the railroad," wrote Thoreau, fifteen years before Russell took his camera out West; "it rides upon us. . . . Some have the pleasure of riding on a rail, others have the misfortune to be ridden upon"—words that prefigure the influential anticapitalist social critic Henry George's 1868 prediction that "the completion of the railroad and the consequent great increase of business and population, will not be a benefit to all of us. . . . As a general rule . . . those who *have,* it will make wealthier; for those who *have not,* it will make it more difficult to get."[105] "The locomotive is a great centralizer," George wrote; "it kills little towns and builds up great cities." The image that comes to my mind is *Coalville*'s desolation row.[106]

Is it surprising, then, that an artist with Russell's obsession was fixated on the icons of crucifixion—an event that, for Christians, at once encoded the wages of sin and the suffering finitude of mortal flesh as well as the eventual permanence of heavenly salvation?[107] Yet—he couldn't help himself—even when gazing at the cross, Russell's eye strayed from the divine to the darkly ironic, and though the telegraph-cross shows up in dozens of Russell's photographs, there's perhaps no other image with humor as black as *East Temple Street, Salt Lake City*, a photo dominated by a telegraph pole towering over a street scene devoted to commerce. A deep look reveals three crosses jutting into the sky (there are three more but Russell has framed the view so that they blend in with

FIGURE 32. A. J. Russell, *East Temple Street, Salt Lake City*, ca. 1869. From *Photographs Taken during Construction of the Union Pacific Railroad*. Yale Collection of Western Americana, Beinecke Rare Book and Manuscript Library.

the tree in the distant background), just as there were three crosses on Golgotha: one for Jesus and two for the thieves crucified alongside him. The three-cross icon would have been familiar to most in the nineteenth century—Currier and Ives's print *The Crucifixion* from 1849 was no doubt nailed to the wall of many a believer's house—and while a lone cross might signify redemption, Golgotha, a landscape, is clearly a place of fear and suffering.

By the late 1860s, "Golgotha" was frequently tied to any landscape of murder, or death, or shame—to any place a person should fear to tread: to the notorious Southern prison camp Andersonville, at which walking Northern skeletons were shot for approaching the "dead line" ringing the camp's interior; to the entire postwar South and its killing fields; to the great cemetery at Arlington; to the arid, bone-strewn Southwest that a Mormon emigrant would have to cross before reaching Salt Lake City.[108] Golgotha, by long-standing tradition, was a hill

FIGURE 33. N. Currier, *The Crucifixion / La Crucificazion / La Crucifixion* (New York: N. Currier, 1849). Library of Congress, Prints & Photographs Division, LC-DIG-pga-03628.

outside of Jerusalem, shaped like a skull.[109] Russell, in 1868, took a photograph of a rocky hill in Wyoming. He called it *Skull Rock*.

9

And so we're back at the beginning, contemplating crucifixion, wondering, what is it that *1000 Mile Tree* was witness to on that day in 1869? Why a sacrifice? And why were we brought to Golgotha?

The great power of *1000 Mile Tree* is that Russell provides no clear answer. He resisted didacticism and instead made a photograph that continues to enchant, from which blows an air of ultimate mysteriousness, "a mad image, chafed by reality," as Roland Barthes put it; but madness doesn't mean the picture is mute, that we shouldn't pay attention.[110] And though photographs—*all* photographs—are emphatically not perfect windows into objective reality, though all photos have no choice but to distort the truth just as they bend the light, they

FIGURE 34. A. J. Russell, *Skull Rock (Granite), Sherman Station, Laramie Mountains.* From F. V. Hayden, *Sun Pictures of Rocky Mountain Scenery* (New York: Julius Bien, 1870). Library of Congress, Prints & Photographs Division, LC-USZC4-11375.

do offer us a path, albeit tangled, through the world's confusion. It's a madman's path, or maybe the way of a holy fool, for appearances are all that they have to offer; but in showing us a history, they also set that history in motion, for in photographs, the hard boundaries between performance and real life, subject and object, and past, present, and future all begin to blur in the middle distance.[111] What a photograph shows can't be divorced from what it means, from what it is, from how it exists in the world among other images, and if we are to glean anything beyond empirical data from a photo, we must struggle with the images themselves, which have their own desires and actively enact their own forward-moving histories. Photographs don't simply record: they also tell. For better or worse, images help guide us in our world, and some of these images are truer and more just than others—though misrepresentations, "raw histories" as the historian Elizabeth Edwards has put it, they inevitably must be.

For instance, when Russell captured the parallel graded beds of the Central and Union Pacific Railroads in the background of a photo ostensibly focused on the intricate trestlework at Promontory Point, he turned his camera into a witness that testified to waste and political chicanery—those things meant to remain invisible from both the train's window and the triumphal narrative of Progress.[112] There was nothing "free market" about the completion of the transcontinental railroad. Congress initially gave away a four-hundred-foot right of way to the Central and Union Pacific Railroads, as well as ten square miles of land for every mile of track laid and loans of $16,000 to $48,000 dollars per mile, depending on how hilly the terrain was. In 1864, as Union and Confederate soldiers were slaughtering one another by the thousands, the railroads convinced the federal government to patriotically double the amount of land granted them per mile and to restructure the loans as mortgages, reducing the corporations' risk to nearly nothing, and so ensuring the corporations' owners vast profits. Because each corporation was paid—at society's expense—according to how many miles of track it laid, it made sense for each to build past one another, for twice the profit, even though only one of the beds would be used.[113] It was a boondoggle of corporate welfare, and if you gaze into the distance of *Promontory Trestle Works*, you can see the two graded beds, two sets of telegraph poles—the traces of mercenary corporate capitalism written on the land, preserved by the Union Pacific's own hired photographer.

Much of Russell's work often functions as investigative photojournalism, and its power comes from the emphatic statement "this really happened." This blood spilled from this body onto this ground. These once-familiar buildings have been reduced to haphazard piles of bricks, dust to dust. These corporations defrauded you.[114] Such images are certainly bad for morale and business: recall the recent Bush administration's ban on photojournalism that depicted dying or dead American soldiers, body bags or government-issue coffins; or the even more recent attempts to ban undercover journalists from capturing the animal abuses rampant on large industrial concentrated animal feeding operations.

Yet, like the very best photojournalism, Russell's images are also

FIGURE 35. Evidence of fraud. See the two parallel graded beds in the background of the photo. A. J. Russell, *Promontory Trestle Work*, ca. 1869. From *Photograph Albums of Utah, Wyoming, Nebraska, and California* (ca. 1867–1869). Yale Collection of Western Americana, Beinecke Rare Book and Manuscript Library.

interpretations; they are, always and finally, art. This tension between document and interpretation, between fact and meaning never gets resolved. That is what makes Russell's photographs worth getting lost in.[115] There's something political about all this, something immediate that resists the attempt to safely bind the dead in the sanctimonious mantle of higher cause and with mystic chords of memory, something that asks the viewer to contend with the tragedy of Progress on a deeply personal level. When read alongside Russell's wartime photographs of bridges or trains, or the engineer Haupt playacting the part of a general, Russell's photos point to the role that technology and engineering and discipline played in the destruction of both human bodies and the land.[116] How many trees have been turned into crosses? How many became telegraph poles?

If images of the railroad in the West are always in some way traces of Manifest Destiny, then Russell's pictures showcase an aesthetic palette of sublime and picturesque tones that mock the harmonious intertwining of industry and nature, and instead reveal the intimately violent clash between physical reality and the hymn of Progress.[117] "Nearly all railroads are bordered by belts of desolation," wrote John Muir in 1918—he could have been describing one of Russell's photographs.[118]

But we won't see any of this if we focus primarily on Russell's published words. When he wrote back to his hometown newspaper of the events he witnessed at Promontory Point, he exclaimed, "The long coveted opening to the markets of the East has at last been furnished by the genius, enterprise, and public spirit of the United States." And then, he followed up with the nearly religious faith that Eden revealed itself to the American pilgrim alone: "since the settlement of [Utah] rain has increased in proportion to the growth of the country and the spread of agriculture."[119] For many today, this is enough to dismiss Russell as just another agent of American empire.

It is, however, significant that Russell's verbal reportage is boilerplate: "The continental iron band now permanently unites the distant portions of the Republic."[120] That word, *unites*, carried many overtones in 1869, a scant four years after Lee surrendered at Appomattox. Americans North and South were tired of war, tired of surveying the wrecked towns, landscapes, and bodies that survived—tired of death. Thousands of glass-plate negatives taken by Matthew Brady famously languished—few wanted to see what they offered anymore and preferred to forget. Yet Russell was a born image maker, someone for whom words paled beside the power of a picture. He had no choice but to look, and the odd tension between Russell's prose and his art is the tension of the struggle to remain alive in an inhuman, antihuman world: his writing was a desperate attempt to remake what he himself knew to be impossible, to reimagine what had already been inscribed on collodion and written on the land.

With photography, though, there are no second chances, just preserved decisive moments. Russell knew this. He had witnessed the horrors of the Civil War, in which well over half a million people lost their lives, horrors that fundamentally altered not just the political bonds of the nation, but the entire culture of living and the work of dying, as well; he had seen whole landscapes—people, buildings, earth, horses—reduced to ashes, smashed to dust.[121] For many, the post–Civil War years saw doubt replacing optimism: how could anyone wholeheartedly believe in American exceptionalism, or of the mastery of nature, after witnessing four long years of reckless murder? Russell had been an integral part of a war strategy that relied on railroads, on capturing them, destroying them, protecting them, rebuilding them. It was a strategy of dispersal that spread the effects of the war up and down the line, leaving thousands uprooted and looking for a new place to settle, a multitude for whom the postwar West seemed to be providential and pure, untainted by fraternal warfare.[122] If the West was a shining future and the East a better-forgotten past, then it was the railroad's unbreakable iron bands that bridged the gap and symbolically healed the rift between North and South.[123] "The annihilation of space and time"—there are few phrases that appear with such regularity in the age of mechanical Progress and Manifest Destiny. But Alexander Pope's original couplet ran: "Ye Gods! annihilate but Space and Time, / And make two lovers happy"; this was the utopian promise of the train.[124] It must have sounded sweet in the years of the postwar United States. And so I hear a desperate note in Russell's all-too-chipper letter home—a letter that would have been read by veterans, and the parents, friends, and acquaintances of those who would never make it back to Nunda—an overly cheerful insistence that the continent had been permanently united, that yes, of course the sacrifices were worth it—they *had* to have been worth it—that technology could make two lovers finally happy; he needed to believe that there was hope in the western landscape, and his hope took a generic, though not disingenuous, written form; but he couldn't buck his muse (who among us can?), and so he turned an ambivalent, monocular eye upon the most modern aspect of Manifest Destiny.

This is why we should pay attention to Russell's photographs: he gives us an environmental visual aesthetic of ambivalence and, ultimately, humility. It's beyond the standard Ansel Adams wilderness vision of a grand, humanless, nationalist nature that graces so many environmentally oriented magazines and websites.[125] It's also different from the tradition of environmental documentary photography: the exploding oil rigs or pesticide plants, the eroded farm fields or refuse-choked waterways in which humans are only ever either victims or destroyers. These are the mainstream environmental visual aesthetics that we have in the contemporary West, and it's hard to overstate their cultural and political importance. Yet in a way, they are both too easy: they don't ask much of us other than to be either dazzled or horrified, and to send a check to the proper organization.[126]

Russell, originally hired merely to document the role modern technology and engineering played in winning the Civil War, couldn't look away, and the chorus made of *1000 Mile Tree* and its accompanying images asks us not to, either—not to pretend that the annihilation of space and time has been an unchecked convenience. There's implied critique in this position, surely, but there is also hope, for Russell's photos invite us to dwell, to linger for a moment with our eyes, to pay attention to *our* world. In the end, they ask us to ponder for ourselves the untold beauties—even among tragedy—that a camera once caught.

10

And what of the tree itself? Thomas Stevens, pedaling his newfangled bicycle over the Rocky Mountains and across the Great Plains, rode beneath its overhanging branches in 1885:

> This tree is having a tough struggle for its life these days. It looks sick and dejected; and one side of its honored trunk is smitten as with leprosy. The fate of the Thousand-Mile Tree is sealed. It is unfortunate in being the most conspicuous target on the line for the fe-ro-ci-ous youth who comes West with a revolver in his pocket and shoots at things from the car-window. Judging from the amount of cold lead contained in that side of its venerable trunk next to the railway few of these thoughtless marksmen go past without honoring it with a shot.[127]

The overexcited, gun-toting passengers, eager to live out their fantasies of Buffalo Bill's Wild West shoot-'em-ups, had been blazing away at the tree for a few years by the time Stevens got there, and they had even destroyed a funereal plate that a grieving husband had attached to it in 1881 to commemorate his wife. She had passed away just as their train chuffed by the rooted memorial.[128] Then, in June 1898, Canadian railway photographer H. C. Barley took the tree's death portrait. Its scourged bark has peeled off in sheets, and it looks like it has been burned; perhaps a stray spark from a locomotive, hurrying its way past Golgotha, landed on a pocket of resin, turning the Thousand-Mile Tree, for just a few minutes, into a burning bush.

FIGURE 36. H. C. Barley, *1000 Mile Tree*, 1898. Yukon Archives, H. C. Barley Fonds, #5485.

ACT FOUR POSSESSION IN THE LAND OF SEQUOYAH, GENERAL SHERMAN, AND KARL MARX

> There is a voiceless beauty we adore
> because we think it innocent and pure.
> We call it Nature, praise its quiet soul
> and save vast tracts from new development.
> But Nature's haunted too, though we don't know
> the names it would be called by,
> the gods it would prefer we reverenced.
> There is no purity. Not in this world.
>
> DAVID MASON, *Ludlow*[1]

1

A pure breath, cold and pristine, sweeps from rocky peaks, tumbling through treetops so high above the ground that a soft susurrus is the only sound descending down to the roots. This is the near silence of the giant cinnamon-red sequoias, trees surrounded by a rainbow of others (gray and yellow pines, black oaks and blue, luminous white firs), trees whose trunks are dozens of feet in circumference, whose bark furrows with age and exposure deeply enough to enclose a human hand in skin meant to protect from fire living wood that pulses blood red and soft pink. One can lie flat, and, if the breeze above pushes hard enough, feel the earth rock with the tree slowly swaying in time to hush-a-bye whispers.

California's Mt. Whitney, standing at 14,505 feet, is the tallest thing in the contiguous United States, and as the climbing eastern morning sun hits its flanks, the mountain casts a long shadowy finger due west, which alights on Whitney's complement, the giant sequoia known as General Sherman—the biggest tree in the world, protected since 1890 by Sequoia National Park.[2] It's hard to avoid the conclusion, in the twenty-first century as in the mid-nineteenth, that California is home to landscapes of binary superlatives: the tallest mountains and, in Death Valley, the lowest point in the United States; temperate rainforests and the country's hottest desert; some of the nation's most productive agricultural land as well as soil so inhospitable that even the prickly Joshua trees are armed to prevent theft.

It's a place well suited to daydreams, especially in the Sierra groves of sequoias, a place where Lilliputian humans unfold their faces sunward—strange, unique leaves—straining for a glimpse of tree crowns growing into the clouds. Ever since America began imagining itself as a continental empire, California has seemed a fantastic promise and an answer, from the 1849 gold rushers to today's silicon-dreaming dot-commers. In 1867, the travel writer and former Union spy Albert D. Richardson wrote of his midcentury rambles throughout the United States: "In exhaustlessness and variety of resources, no other country on the globe equals ours.... Its mines, forest and prairies await the capitalist. Its dusky races, earth-monuments and ancient cities importune the antiquarian. Its cataracts, canyons and crests woo the painter. Its mountains, mineral and stupendous vegetable productions challenge the naturalist."[3] Though he was writing about all the land west of the Mississippi, Richardson might as well have been penning a love letter to the synecdochic Golden State: the frontispiece to his *Beyond the Mississippi* (1867) is a California dreamscape of fantastic proportions, and the entire lower half of the image is meant to suggest the Yosemite Valley with its commanding El Capitan off on the left facing the famous Bridalveil Falls on the right. Occupying center stage of the fantastic four-ring circus is an imagined train, the headlight of progress scattering wild Indians, who

POSSESSION IN THE LAND OF SEQUOYAH, GENERAL SHERMAN, AND KARL MARX   163

FIGURE 37. From Albert D. Richardson, *Beyond the Mississippi: From the Great River to the Great Ocean* (Hartford: American Publishing Co., 1867).

flee, frightened, to the safety of the woods. To the left and right of the main attraction, farmers reap their rich rewards, while miners chip away at a hillside and towering trees on left and right look on. Gold and silver, territory, forests, coastline, markets, deserts blooming like roses: "the original Garden of Eden," as one writer put it in 1864.[4]

Once I walked the sequoia's range, much of it preserved, from

Yosemite — near where the landscape history of the sequoias begins — on south through miles of wilderness to Sequoia National Park, this essay's terminus, all to sit on the summit of the United States and watch the sun rise. It's a landscape whose elemental beauty is hard to credit: empty and peaceful and quiet, there's a crystalline quality to the air that fills both the lungs and the eyes with light. Yet every one of those miles I wandered was through land that was once someone's home, an Eden in which the silence and violence of expulsion has been extended indefinitely.

And this is the problem: how do you tell a history of silence?

2

One way to begin is with the beginning, in the north; and in this beginning, there was gold.

California became world famous in 1848 for the discovery of golden flakes near Sutter's Mills, an event that Karl Marx called the most important event in American history.[5] But shortly thereafter, fame of another sort followed. In 1852 word began to spread from Calaveras County, just south of the Rush's epicenter and not far from present-day Yosemite National Park, of trees so immensely large that even those familiar with the coast redwoods — trees capable of growing more than three hundred feet in a span of 1,500 years — stood stunned in their shade.[6] The Western world had known of the redwood since the late eighteenth century, and its first detailed description appeared in print in 1776, by Spanish missionary Fray Pedro Font.[7] But it wasn't until Augustus T. Dowd, gun slung over his shoulder, strode back to his hunting camp with tales of a giant grizzly, one bigger than anyone had ever seen, that the West began to learn of the Big Trees. Dowd was a gunner who supplied meat to the workers busily digging the lifeblood canals for John "The Pathfinder" Frémont's Mariposa gold-mining operation, and on that day in 1852, so the story goes, Dowd had been tracking a wounded grizzly when he stumbled into the Calaveras Grove of giant sequoias; nose to the trail, he "suddenly came upon one of those immense trees that have since become so justly celebrated throughout the civilized world," wrote one of the Sierra's first boosters, former miner J. M. Hutchings.[8] "All thoughts of hunting were absorbed and lost in the wonder and surprise inspired by the scene. 'Surely,' he mused, 'this must be some curiously delusive dream!'"[9]

There had been previous tales of the mammoth things—dismissed as mere rumor—but when Dowd led his incredulous friends to his prize, the world suddenly craned its collective eye skyward, if only for a moment.[10] For Dowd's tree, the Discovery Tree, was promptly cut down, the prostrate trunk fashioned into a bowling alley, the stump into a dance floor capable of accommodating thirty waltzing couples. "However incredible it may appear," wrote Hutchings, who loved to kick up his heels, "on July 4, 1854, the writer formed one of a cotillion party of thirty-two persons, dancing upon this stump; in addition to which musicians and lookers-on numbered seventeen, making a total of forty-nine occupants of its surface at one time!"[11]

And a man named George Gale, an Argonaut frustrated by his inability to get rich enough quickly enough with pick and shovel, saw opportunity. Into the Calaveras Grove of Big Trees he went, looking for the largest trophy he could find, until his roving eye settled on the Mother of the Forest. Industriously, methodically slicing deeply into her bark, he slowly skinned back eight-foot sections one at a time to expose her soft inner wood, finally stripping her to the height of 116 feet until she stood rose red and gleaming wet, utterly destroyed. His work done, he sent his trophy east, to the Crystal Palace in New York City, site of the Exhibition of the Industry of All Nations—the World's Fair of 1853—presided over by the Prince of Humbug himself, Phineas T. Barnum, the impresario who put on display the mermaid from Fiji (a creature with the head of a monkey and the tail of a fish), and General Tom Thumb, the man who never grew taller than a six-month-old infant.[12]

It is an accident that the giant sequoia entered American history, and then only through its coincidental, enduring association with gold. That association is a big part of the reason that it took so long for Euro-Americans to find them in the first place: the Sierras are rugged, and the winters are famous for dropping wet loads of snow ten, twelve, or more feet deep. Unlike the redwoods, the sequoias stand aloof in inaccessible places, scattered widely along an extremely limited range from Placer County, near Reno, south to present day Sequoia National Park—a distance of only 260 or so miles—rooted in a narrow band between five thousand and eight thousand feet of altitude that is nowhere more than 15 miles wide. Finally, these shy trees keep to their own: they are social beings that huddle together for protection and company in groves, and depending on how one defines "grove," there are only between sixty-five and seventy-five sequoian communities that currently exist. Loners are rare. Because Spanish and then Mexican settlement tended to take advantage of the agricultural and forest lands along the coast, there was little incentive to wander inland to the inhospitable peaks of the Sierras until the feverish pursuit of gold attracted Americans to California's most rugged areas, flushing the sylvan beauties from their cover.

If there was gold in the hills, by the 1850s a few enterprising Americans realized that financial gain could also be found arcing high overhead. The stump-dancing J. M. Hutchings first visited the Yosemite Valley in 1855, and he began promoting it—along with his hotel located on the valley's floor—the minute he returned: it was Hutchings (who would one day employ John Muir as his property manager) more than anyone else who put Yosemite and its trees on the tourists' map.[13]

At nearly the same time, a young New Englander named Galen Clark found himself strolling through the 1853 World's Fair grounds, and though he just missed seeing the Mother of the Forest's flayed hide, he witnessed enough Californian gold dust on display to whet his appetite for wanderlust and social mobility. Soon after, Clark made his way to the Sierras, where, like A. T. Dowd before him, he was drawn by the magnetism of Frémont's Mariposa. Initially a Spanish land grant idyllically named for butterflies, Las Mariposas was snared by Frémont just as gold was discovered at the end of the Mexican-American War, and the Pathfinder's property—seventy square miles sitting right in the middle of the mother lode, which came to include six mines, two towns, a railroad, and a rent-paying population of more than seven thousand souls—became the "jumping off place of civilization" for tourists interested in seeing Yosemite Valley or the Big Trees.[14] Las Mariposas had a mine named Pine Tree and a peak named Mt. Bullion and a business plan hoping to make it all "bleed at every vein," as the pastor Thomas Starr King put it when he toured Frémont's works in the winter of 1860–1861; but Clark quickly tired of laboring for another: he wanted trees and mountains of his own, and it didn't take him long to discover Yosemite.[15]

It was an auspicious moment to be dissatisfied with grubbing: in 1851 Major James D. Savage and his Mariposa battalion had just finished their job of ensuring that nothing human stood between the golden nugget and capitalism's invisible, forever-grasping hand. As forty-niners flooded into the Yosemite area, they found that the land was already home to many Native peoples whom we today all lump together under the general classification "Interior Miwok."[16] In an act of typological irony, the miners disdainfully called the American Indians they found "Diggers"—a racially charged swipe at the Indians' practices of gathering roots for sustenance—but the Diggers called themselves the Pohoneechee, Potoencie, Wiltucumnee, Nootchoo, Chowchilla, Hownache, Mewoo, Chookchance, and Ahwahneechee.[17] It was this last group, the Ahwahneechee, that has become most associated with Yosemite Valley, and Clark tells us in his short reminiscence from 1904, *Indians of the Yosemite Valley and Vicinity*, that by the time the forty-niners arrived, the Ahwahneechee had already been decimated by three centuries of disease and war wrought by the geopolitically destabilizing entrance of first Spanish, then Mexican, and finally American imperialists.[18] But it wasn't until the late 1840s and early 1850s—when miners began swarming over Indian lands, destroying the oak trees whose acorns were a staple of the Ahwahneechee diet, killing off the game, forcing the American Indians into the mines, and, for those miners hailing from the American South, turning them into slaves—that the Sierra Indians felt themselves at a point of crisis.[19] The gold rush turned out to be a turning point for California's Indian populations; their numbers plummeted precipitously—by around 80 percent—as a direct result of the white race for golden riches.[20] When the Ahwahneechee and other nearby groups started to resist the miners' depredations, the US government began a policy of forced relocation and murder. In a story whose cadences are depressingly, repetitively well worn, the Ahwahneechee fled to the protective fastness of an inhospitable environment (in this case, the Yosemite Valley), wild-eyed rumors started circulating of the atrocities committed by these off-the-reservation American Indians, and before long, the newly formed Mariposa battalion's bugle echoed from the hills of the Sierras. "Active preparations were accordingly made by the State authorities to follow [the Ahwahneechee], and either capture or exterminate all the tribes involved," Clark tells us.[21]

This is how Yosemite was discovered: Savage's butterflies mounted their horses, hoisted their well-oiled rifles, and flew to the Ahwahneechee valley, whereupon they found themselves dumbstruck by the towering granite domes, upthrust needles, and lush meadows of the Yosemite. Volunteer Lafayette Houghton Bunnell, one of the mythical first white men of discovery, later wrote that "none but those who have visited this most wonderful valley, can even imagine the feelings with which I looked upon the view that was there presented . . . as I looked, a peculiar exalted sensation seemed to fill my whole being, and found my eyes in tears with emotion."[22] But Bunnell didn't have long to play the tourist, and his reverie was interrupted by Major Savage: "'You had better wake up from that dream up there, or you may lose your hair. . . . We had better be moving; some of the murdering devils may be lurking along this trail to pick off stragglers.'"[23] And so the Mariposa battalion set to work exterminating the Ahwahneechee.

Savage's men did their job well: "There is a report," wrote one correspondent, that the soldiers "defeated the Indians, killed three hundred, and [had] taken one hundred and fifty squaws."[24] Bunnell and his comrades efficiently almost erased the American Indian presence from Yosemite—though various bands of American Indians continued to fight incursion into their land—and in so doing gave us the landscape that has become so well known, so well photographed, so revered as one of the crown jewels of the US environmental movement.[25] But Savage's men didn't—perhaps couldn't—excise everything native, and Bunnell, who was as handy with a pen as a gun, once again found himself fantasizing beyond the here and now as he and his fellow soldiers huddled around a campfire, trying to put words to their experiences, trying to give the landscape they could barely believe a name. After considering and rejecting first one, then another, Bunnell struck gold: "'I ... proposed that we give the valley the name Yo-sem-i-ty, as it was suggestive, euphonious, and certainly *American*; that by so doing, the name of the tribe of Indians which we met leaving their homes in this valley, perhaps never to return, would be perpetuated.'"[26]

Adirondack mountain echoes in the Yosemite Valley: *Yo-sem-i-ty*, thought Bunnell, meant "grizzly"; he thought it was what the Ahwahneechee called themselves, the Grizzlies—though it seems more likely that *Yosemite*, an Anglicized bastardization of *johemite*, was a western Miwok pejorative for the Ahwahneechee meaning "some of them are killers." In any case, it does make a certain twisted sense: Bunnell and the rest of the battalion derogatorily referred to the Ahwahneechee and all American Indians in general, as "grizzlies"—fearsome, of the West, and fun to kill.

It was here, in this newly wrought wilderness, that a dissatisfied Galen Clark found the tree of life at the center of his own Eden, when, in 1855, he began chasing the rumors of a hunter named Hogg who had seen a few giant trees while rooting around in the area. A year later, Clark burst into what he named the Mariposa Grove of Big Trees, home of one of the most famous sequoias—the Grizzly Giant.[27] Soon after, Clark, like Hutchings, opened his own eponymous hotel, Clark's, the first major destination for Big Trees tourists, and he inaugurated a swift tourist business, a temporary home to tree-crazy luminaries including Ralph Waldo Emerson, John Muir, and the century's most famous American natural scientist, Asa Gray.[28]

What began in the north continued, with remarkable fidelity, in the south, and in 1858, yet another miner-cum-provisioner, Hale Tharp, was shown a grove of Big Trees that would eventually be recognized as the most extensive continuous forest containing the largest specimens of sequoias in the Sierras, trees that would steal some of Mariposa's limelight. Tharp was part of the secondary wave of Argonauts who, once the areas around the initial discovery of gold at Sutter's Mills were tapped out, followed the Sierras south into an increasingly rugged territory, looking for promising prospects. In 1856, Tharp settled along the banks of the Kaweah River, and though he is often credited as the first white man to take up permanent residence in the Kaweah area, the landscape had been home to various groups of Monache (Balwisha, Waksachi, Wobunuch), Yokuts (Yaudanchi, Wukchamni, Gaiwa, Yokod), and sometimes Tubatulabal Indians for at least five hundred years.[29] What Tharp found was a landscape teeming with life: "There were about 2,000 Indians then living along the Kaweah River," he recalled in 1910. "Deer were everywhere, with lots of bears along the rivers, and occasionally a grizzly bear. Lions, wolves, and foxes were plentiful. There were a great many ground squirrels, cottontail, and jackrabbits; quail were seen in coveys of thousands.... There were plenty of fish in the rivers."[30] This wilderness garden was assiduously maintained by its native inhabitants, who had kept the woodlands open, both by gathering fuel for their fires and by occasionally setting small blazes to control the underbrush. But they didn't farm—they were nomadic hunters, fishers, and gatherers—and so Tharp's eyes would have recognized nothing that looked settled.[31] The high Sierra passes meant that various bands from the eastern and western sides of the Sierras could trade with one another, and the Kaweah region saw a flourishing network of American Indian commerce develop: pine nuts were exchanged for acorns, obsidian for shell beads, rabbit skins for deer hides.[32]

But there is something else that neither Tharp nor the standard accounts of the sequoia region tell us: this Eden had been ground zero for the Tule Indian War of 1856—the very year that Tharp set up camp along the Kaweah River.[33] Like Yosemite before it, hostilities had erupted after rumors of Indian breakouts, rumors that, even at the time, were shown to be entirely fanciful. Rightfully fearing massacre, various bands of American Indians fled for the hills, which only confirmed white suspicions of treachery. Soon enough, a company of volunteers and enlisted men four hundred strong set out in pursuit of their quarry, dragging behind them a howitzer, and, meeting with fierce resistance, evidently decided that extermination was the only humane course of action. Not a single white person was killed in the decisive battle, though one hundred American Indians were, and their physical claim to the landscape was effectively extinguished.[34]

Two years after the army did the only thing it does, Tharp was invited by one of the survivors, an Indian he called Chief Chappo, to see the sights and was led to what we now know as the Giant Forest—so named by John Muir in 1875—home of the largest trees in the world. The American Indians in this region were mobile cosmopolites: they had seen, heard, and witnessed firsthand the depredations that followed hard on the heels of the first white man, so when Chappo showed Tharp the Giant Forest, it's a good bet that he was doing so as part of a complex strategy of survival, one that included preemptive appeasement, in an effort to stave off genocide.[35]

It was here, near the extreme southern limit of the sequoias' reach, that the biggest specimen of the world's largest tree stood, looking north, surveying a landscape of Indian wars, the race for wealth, burgeoning tourism and incipient environmentalism, and the competing imperial dreams of the world's great Western powers. In 1879 this tree would be named General Sherman.

3

Another way to begin, though it changes the story, is with the tree.

There is a basic problem in trying to write an essay of the sequoian landscape, and the problem is this: how does one affectively chronicle the miraculous?[36] For, in the mid-nineteenth century, gazing upon a sequoia was to see something unprecedented, something impossible, something for which previous experience could only leaving one stunningly ill equipped. The sequoias were simply unthinkable—at least to Euro-American minds; but by the 1870s, when their painted and photographed forms saturated American culture, the possibility of an authentic, unmediated experience was lost.

We will never know what it was like to experience a sequoia for the first time.

I have to imagine that Clark—whose prose and poetry outs him as a romantic dreamer—was stunned, elated, confused, and ultimately reborn. He had been a painter before he came west, and, by the mid-

1850s, was a consumptive. One of the reasons he decided to live among the Big Trees was for their curative powers.³⁷ He must have looked down at the sun-dappled, pungent dried duff surrounding the trees, and gaped upward, thinking about how long it would take a single thornlike needle to fall from the tree's crown. Tharp, too, must have tilted his head far back enough for his skull to rest in the yoke of his shoulder blades, squinted into the sun, and, for a moment, stood very quietly. I want to believe that he, too, marveled, and that making for his first home a cabin inside the trunk of a prostrate sequoia was an act of poetry as well as one of history: Tharp literally became a dryad, the tree-living nymph from Greek mythology.

What was it like to experience a sequoia for the first time?

John Muir tells us, with his characteristic purple prose, that it was like walking into a medieval cathedral: "When I entered this sublime wilderness the day was nearly done, the trees with rosy, glowing countenances seemed to be hushed and thoughtful, as if waiting in conscious religious dependence on the sun, and one naturally walked softly and awe-stricken among them. I wandered on, meeting nobler trees where all were noble, subdued in the general calm, as if in some vast hall pervaded by the deepest sanctities and solemnities that sway human souls."³⁸ Fitz-Hugh Ludlow, the famous travel writer and hash eater, had his mind blown:

> Before reaching Clark's we had been astonished at the dimensions of the ordinary pines and firs. . . . But these were in their turn dwarfed by the Big Trees proper, as thoroughly as themselves would have dwarfed a common Green-Mountain forest. I find no one [in the East] who believes the literal truth which travellers tell about these marvellous giants. People sometime think they do, but that is only because they fail to realize the proposition. They have no concrete idea of how the asserted proportions looks. . . . I freely confess, that though I always thought I *had* believed travellers in their recitals on this subject, when I saw the trees I found I had bargained to credit no such story as that, and for a moment felt half-reproachful towards the friends who has cheated me of my faith under a misapprehension.³⁹

And Mary H. Wills, who penned one of the earliest responses to the sequoias in the Mariposa Grove by a woman, turned her back on descriptive language in favor of the statistical: "I saw them ninety-nine or thirty-three yards in circumference, and sat in the stage drawn by four horses through one which only measures eighty-four feet. They are from one hundred and fifty to two hundred feet high, and show from their concentric rings or layers that they are four thousand years old."[40]

Muir, Ludlow, and Wills, like Hutchings and Clark before them, were all seeking to promote the Big Trees for various overlapping reasons — preservation, tourism, literary fame — and the usual scholarly analysis is that these writers, along with a host of other essayists, promoters, photographers, and artists, "frame the view" for us, that their work structures all later experience of the things.[41] There is good historical reason for the framed-view interpretation, for looking at the landscape, like looking at art, has often been an exercise in cultivation: appreciating beauty the right way is one marker of class distinction. But what the frame's definite edges miss is that each writer finds her or himself groping in the dark for illumination, stretching the boundaries of metaphorical, allusive language to convey some sort of authentic experience; yet in doing so, each winds up missing the trees for him- or herself. The sequoias are a sacred nave, or a Vermont forest shrunk to blades of grass, or abstract numbers, ninety-nine feet and thirty-three yards. They're anything but sequoias. The only way that Thomas Starr King could begin to make sense of the mammoth things was to transport them away from their native surroundings, to compare them to the biggest things he knew: the Great Elm on Boston's Common, Trinity Church in New York, or the monument on Bunker Hill. Any of these beside a giant sequoia would be like "General Tom Thumb at the knee of Hercules," the fantastic beside the imaginary.[42] It seems that the trees had an almost preternatural power of transmutation: if you had been able to ask a Californian living in the Sierras in the 1850s what a grizzly was, he might respond that it was a bear, or a tree, or an Indian; and when the boundaries between discrete things become as blurry as they do among the sequoias, when the only way to describe a thing is through metaphor, then a void in meaning inevitably yawns open, breaking the frame and clearing the stage for a dream-play of images, words, meanings. Ludlow is the best guide: "When I saw the trees I found I had bargained to credit no such story." Words simply failed among the Big Trees.

As it turns out, so did vision, and perhaps the most famous image of a sequoia is photographer Carleton Watkins's *Grizzly Giant, Mariposa Grove*, a photograph self-evidently stretching the limited possibilities of its own conception.[43] On first glance, the Grizzly Giant doesn't look *that* big, but the tiny characters at its base, so small that they almost disappear into bark and collodion, tell us otherwise. That's wonderful, but stranger yet is the bulge in the tree as it stretches skyward, the convex warping in the middle of the frame, a distortion induced by the camera's lens; and then there's the glaring tonal contrast—just right at the tree's base but washed out above, the range of real light, from deep shadow to brilliant-blue sky, too broad for the mechanical eye's narrow window of exposure, which loses some of the definition of the Grizzly Giant's crown, instead dissolving it, ghostlike, into the oblivious horizon.

If Watkins's *Grizzly Giant, Mariposa Grove* is the most famous photograph of a sequoia, then Albert Bierstadt's *The Great Trees, Mariposa Grove* (1876) is certainly one of the more widely circulated paintings— and clearly based on Watkins's earlier work. It's also the biggest: at ten feet by five feet, *The Great Trees* mimes the spatial impact of the sequoias. Originally exhibited at the 1876 Centennial International Exhibition of Arts, Manufactures, and Products of the Soil and Mine— yet another World's Fair appearance for the sequoia—Bierstadt's painting was shown alongside a who's who of famous American artists and their work in a celebratory reckoning of how far the distinctive visual aesthetic of Nature's Nation had come since the republic's founding, though many of Bierstadt's visual cues were repetitive and well-worn by 1876: the tiny people in the foreground, the surrounding trees needling upward, the dramatic lighting. *The Great Trees* was skewered by the critical press for being too obvious—even though Bierstadt was given an award of eminence by the centennial commission (so much for both critical judgment and awards).[44]

Yet among all that is solid rises an airy, invisible giant just to the right

FIGURE 38. Carleton E. Watkins, Sequoia Gigantea—*Grizzly Giant—Mariposa Grove*, ca. 1866. Albumen silver print 20½ × 15 in. (52.1 × 38.1 cm). The J. Paul Getty Museum, Los Angeles. Digital image courtesy of the Getty's Open Content Program.

FIGURE 39. Albert Bierstadt, *The Great Trees, Mariposa Grove*, 1876.

of the Grizzly, the negative space of its trunk exactly mirroring its twin as their branches intertwine. If the painting works, it's because it wanders over a California dreamscape where fantasy gives heft to fact: by placing the great tree in an imagined landscape and giving it the form of a real sequoia—the Grizzly Giant—by rearranging and simplifying space so that it can be apprehended without distortion, and above all by planting it next to a ghostly other, Bierstadt has managed the trick of making the unseeable real.

One reason that neither words nor images could frame the Big Trees was that the Western cultural categories of nature appreciation had been devised in a different landscape, tutored by different trees. William Gilpin's theorization of the picturesque depended on a pleasantly wooded landscape; indeed, on the first page of *Remarks on Forest Scenery* he wrote, "it is no exaggerated praise to call a tree the grandest and most beautiful of all the productions of the earth," and his *Remarks on Forest Scenery* is at once a field guide, a philosophical argument, a landscape manual, and a love letter to pleasing British arboreal forms.[45] But Gilpin's picturesque evolved in a pastoral English landscape, one that, through centuries of intensive agricultural land use, had been scaled to a human size, and when it eventually migrated to the sequoias, it found itself in alien territory. Perhaps the sublime was a better fit—yet the sequoias were a bit too sedate, too rooted, and the only pain one was likely to feel taking in the vast prospect of a sequoia's trunk was from a pinched nerve in the neck. The Big Trees were simply too new, too unlike anything the West had ever encountered. Or, as Ludlow put it, "The marvellous of size does not go into gilt frames. You paint a Big Tree, and it only looks like a common tree in a cramped coffin."[46]

This failure to safely frame the view of the sequoias with word or image often bred discontent, and for every Muir-esque description of the sequoias that takes off on dizzying flights of hyperbolic metaphor, there is another pouting feeling that "the trees do not look as you expected; they are not as large; 'they look as if somebody has stripped off their clothing, and left them in nightdress.'"[47] Even the normally chipper Thomas Starr King found himself disappointed when looking at his first sequoia: he supposed it a sapling, merely, no more than forty feet around, even though such a wispy shoot would still be twice as large as Boston's famous elm.[48]

Paradoxically invisible and gigantic, the sequoia groves ultimately allowed the possibility for dreaming, and while the George Gales saw no further than the wealth that might be stripped from a sequoia, and the Frémonts mucked about on hands and knees "engaged in extracting the treasure from the chinks and pockets of Mother Earth," others, contemporaries of Gale and the Pathfinder, found themselves haunted by the implications of a tree over three hundred feet tall and nearly two dozen in diameter.[49] "What silence and mystery," wrote King, who considered himself one of those sensitive souls upon whom nature could make its mark. "How many centuries of summer has such evening splendor burnished thus the summit of the completed shaft," he mused while lying underneath a particularly large sequoia in November 1860. "How long since the quickening sunbeam fell upon the first spear of green in which the prophecy of the superb obelisk was enfolded? Why cannot the dumb column now be confidential . . . why will not the old patriarch take advantage of the ripple through his leaves and whisper to me his age? Are you as old as Noah? Do you span the centuries as far as Moses? Can you remember the time of Solomon?"[50]

4

Or, perhaps this is the best way to start: A contest of meanings rushed into the void created by the Big Trees' frame-shattering newness, all competing to make sense of the limbs, and crowns, and deeply burrowing roots. Financial gain was on many a mind, but surprisingly few advocated for outright logging—at least initially. As the nature tourism industry came of age in the later half of the nineteenth century, the Big Trees became one of the must-see sights on any grand tour of the United States, and there was little trouble envisioning them as a natural, sustainable source of wealth. By the 1870s and 1880s, countless photographers had sold thousands of images to the public, and if money didn't exactly grow from sequoian limbs, tourist dollars certainly sprouted in their shade—as long as the axes were kept at bay.[51]

But there's a more complex narrative that begins to unfold as soon as Hutchings publicized Dowd's discovery: these trees were seen as undoubtedly American, the equivalents of the Egyptian pyramids and Roman ruins, national monuments conferring the sort of ancient cultural capital that would allow the United States to hold its head high among other modern, civilized nations.

The trick that the sequoias had to perform, though, was that California was absolutely not an American space, as everybody knew: until the gold rush, California's Indians far outnumbered non-Indian populations.[52] And for the three centuries preceding the 1848 Treaty of Guadalupe Hidalgo, the land had been claimed first by Spain and then Mexico; Spanish names—San Francisco, Los Angeles, the Sierra Nevada, even California itself—dotted the landscape. Toponyms aside, California was an intensely diverse place, too: the rush for easy wealth meant that the Harvard-educated sons of blueblood Boston Brahmin scrabbled about in mines alongside Chileans, Mexicans, American Indians, African Americans, Chinese, the middling and poor, and many others. Finally, California was the bull's-eye of Western geopolitics, the point of intersection for Western imperial desire. Besides Spain and Mexico, Russia had planted its nation's flag on Californian soil in 1812 when the Ross Colony was established by Russian merchants who sent Aleutian Island and Kodiak Island Indians to hunt for sea otters, farm the land, and even begin industrial production, all in service to the mother country.[53] So it took an almost perverse will to assert that California was American in any sense at all.

It was this historical, multicultural context that nationalists in the United States tried to make the sequoias efface, and the most common strategy was to tell a story of the trees as transhistorically American.[54] If the past was only ever a prologue setting the scene for today's main act, and if the Big Trees were the long-lived prophets of empire's westward course, then the prickly identity crisis could be avoided: nations, tribes, empires might come and go, but the trees remained; they alone were timelessly American. And because they were, according to the script, outside of history, there was no need for tree time to flow linearly: the Big Trees could seamlessly connect the present with a long-gone history, leaping over the contested middle ground of Indian, Spanish, Mexican, and Russian possession, and so testify that California had been consecrated American long before a single white foot ever touched a square inch of North American soil.⁵⁵

So it's a great irony that the sequoias were almost named for an English war hero when, in 1853, English botanist John Lindley, having received a few samples of cones, foliage, and wood, named the new trees the *Wellingtonia gigantea* after the British general who bested Napoleon at Waterloo. Few Americans outside of Asa Gray's Harvard University took notice of Lindley's appropriation, but luckily for US nationalism, the following year, while George Gale and others got down to the work of making the Big Trees pay, the French botanist Joseph Descaine decided that *Wellingtonia gigantea* was not, in fact, an entirely new thing under the sun; it was actually related to the redwood, he argued, a tree named by the Austrian linguist and botanist Stephen Endlicher, *Sequoia sempervivens*. And so Descaine, the Frenchman, decided to follow the Austrian's lead and give the Big Trees a Native American name: *Sequoia gigantea*. When Gray and his mentor, John Torrey, began popularizing the trees as such, the name stuck.⁵⁶ Sequoia.

The historical record claims that "the genus was named in honor of . . . a Cherokee Indian of mixed blood, better known by his English name of George Guess, who is supposed to have been born about 1770 and who lived in Will's Valley, in the extreme northeastern corner of Alabama, among the Cherokees."⁵⁷ Yet Sequoyah may have been nothing but a ghost dreamed up by a white imagination.

The tale of the "Cadmus of the Cherokee," so nicknamed in honor of the mythological Greek king who devised the first Greek alphabet, first made headlines in the 1820s, and though almost no hard evidence has ever surfaced, the broad outlines of the Sequoyah story have been repeated faithfully and become accepted historical truth.⁵⁸ In the early nineteenth century, Sequoyah or George Guess, or sometimes George Gist, becomes obsessed by "talking leaves," or written notes that he saw

American traders and missionaries passing to one another. Sequoyah is invariably an outsider, marked sometimes by mixed blood (in some versions he is the son of a Cherokee mother and German man), or living on his own as a reclusive hermit, or is lame, or a serious alcoholic, or a derelict father and husband; but he always has one positive trait, usually attributed to his European lineage: a tenacious intellectual curiosity. And so his investigations into the mysteries of the talking leaves became an obsession. Thanks to the charitable work of local Christian missionaries who, depending on the narrator, lend various amounts of help to the native scholar, Sequoyah one day hit upon the logic which allowed him to craft a syllabary of eighty-six characters (or eighty-five, or ninety-two; the details frequently change), whose use rapidly, some might say miraculously, spread among the Cherokee. It was a wondrous invention, "a phenomenon unexampled in modern times" wrote one traveler in 1828, even if it was often understood to be a rude language incapable of the subtle artistry and fine turns of phrase indigenous to English. Inevitably, Sequoyah then became a peaceful mediator between the Cherokee and the federal government, one who negotiated for land in Indian Territory, and then helped usher his people out of the Southeast and along the Trail of Tears. And so the "illiterate Indian genius" became, in the words of one of his earlier biographers, "the only man in history to conceive and perfect in its entirety an alphabet or syllabary."[59] In only a decade.

The problem with this story, which even today goes largely unexamined, is that it far too conveniently fits into the colonial narrative of white superiority; the half-white Sequoyah becomes the exception that proves the rule of savage Indian illiteracy, and the strife that characterized politics of the Southeast in the late eighteenth and early nineteenth centuries—bitter feuds between American Indians and American squatters and politicians, and between factions within the Cherokee nation itself—winds up triumphally overwritten by the spectacle of an Indian who can write and who, of his own volition, sees the wisdom behind forced removal.[60]

Yet in 1971, an American Indian historian named Traveller Bird, who claims direct descent from the actual Sequoyah, published an alternative version of the myth. First, there never was a Sequoyah, because the name is meaningless in Cherokee. The actual person was known as Sogwili, or George Guess, a name Sogwili appropriated from a white raider he caught and killed for stealing Cherokee cattle. There was never a German father in Sogwili's life, either; instead, he was the child of a native man and woman. In the late eighteenth and early nineteenth centuries the Cherokee nation was riven over conflicts between the progressives—who wanted to assimilate into American society—and the conservatives. The conservatives, including Sogwili, fought to preserve their ways, one of which was the Scribe Society, a cadre of powerful Cherokee who had passed the secrets of writing down from one generation to the next since, at least, the fifteenth-century. Lest this history of an exclusive literacy sound fabulous, remember that reading has traditionally been a tool of the powerful, whether it be Catholic masses conducted in a dead language like Latin, or slave owners forbidding their slaves to own books. Presciently fearing that Cherokee lands would be overrun by gold-seeking white settlers—the hills of Georgia and Arkansas were cursed with the metal—Sogwili founded a colony of Cherokee in the Mexican province of Téjas at the end of the eighteenth century.

But Sogwili couldn't leave well enough alone, and he returned constantly to the Southeast, supporting the conservatives who remained, and riling the progressives up. In 1795, Sogwili made the fateful decision to dissolve the Scribe Society, to open the Cherokee language to all conservatives in an anticolonial effort to resist white cultural invasion—a democratizing of the liturgy. For this he was found guilty of witchcraft by a jury of progressives, who cut off his ears, all of his fingers, and branded him on the forehead.

By the 1810s, however, word of Native literacy was starting to leak out: conservatives were writing messages to one another everywhere, even on trees, stymieing missionary efforts to Americanize, and control, the Cherokee. Rather than admit to an indigenous language developed independently of the West's civilizing light, and so implicitly acknowledge Cherokee civilization—the story we have today was devised. The disgraced Sogwili, clearly, could not be made into the Cherokee

FIGURE 40. John T. Bowen, *Se-Quo-Yah* (Philadelphia: F. W. Greenough, 1838). Library of Congress, Prints & Photographs Division, LC-DIG-pga-07569.

Cadmus, so another man, whose name was George Gist—*Gist* was close enough to *Guess* for the purposes of imperialism and history— was given the name Sequoyah, and an invented background of mixed-blood ancestry tacked onto the fanciful name. When it came time to paint Sequoyah's portrait, Gist couldn't be found, and Sogwili had only stubs for fingers, so Thomas Maw, a random, though unmarked Cherokee, was chosen to represent the founder of the language.[61]

There's nothing triumphally American to Traveller Bird's history, and when Sequoyah's name was given to the Big Trees, it was the story of a rude savage bearing white blood who ingeniously crafted a language, with missionary help, before graciously heading west into oblivion that was grafted onto the trees' trunks. Just so did America find itself with a supposedly unique, Native, natural symbol of national, predestined might, whose shape, longevity, and sheer might were all perfect for marking the course of empire.

If you could have asked an American of the middle nineteenth century what or who Sequoia was, what would the answer have been? A tree? A remarkably learned, peaceful, accommodating Indian? An obnoxious renegade? A man of German and Cherokee decent named Gist? A handsome painting of an Indian with long, graceful fingers and unblemished features? The most honest answer is simply this: a man whose last name was cryptically Guess.

5

Perhaps one can credit no such history as Traveller Bird's—it certainly refuses to be framed by historical scholarship's disciplined rules of engagement. But if we dismiss Traveller Bird as ideologically driven, overly romantic, and, in his refusal to cite well-known sources in the proper academic manner, untrustworthy, then we must also wave the tale of the Cherokee Cadmus away as equally tendentious and unsupported.[62] We may never really know who Sequoyah, Sogwili, or George Guess was, but we can essay to reveal the cultural politics of a celebrated species of tree with the name sequoia. There is something troublingly bizarre at work in the Big Trees: at the very same time that American Indians were being driven from their lands and hunted down, during the very same decades when tales of Indian raids and the punitive massacres that followed hard on their heels filled the press, during the same era when the post–Civil War US government turned its full military weight toward "pacification," the sequoias were being celebrated as American because they were Indian.

Perhaps the tale told by the world's largest witness tree is the bitterly ironic story of a story, one spun of murder, dispossession, appropriation, and historical fabrication. Irony is a weapon: one of its great powers is to slice deeply into contradiction, to expose the absurdity of the narratives that fill our world.[63]

And so all the gold, produce, and lumber cut out of California's landscape was dug, grown, and felled by someone, and that someone was often Native: American Indian labor built California.[64] Long before the Joads ever picked a poorly paid pea, one of the Golden State's chief advantages for capitalists was that labor costs were kept artificially dirt cheap, and Native workers were often paid nothing more than food and clothes—if they were paid at all: before, and even after the Civil War, the profits from Indian abduction and slavery bought many a fine suit for America's free marketeers. After the Mexican-American War, California's Indians rapidly lost their demographic dominance and with it, their geopolitical might. Those that survived war and disease found ready work, if not wages, in the mines, as domestic servants, and especially, as agricultural laborers—until, that is, white Euro-American laborers saw American Indians as undercutting their wages. What ended up developing, whether by intention or improvisation, was the creation of a class of permanently migratory, permanently disenfranchised people who could either be lassoed into the labor pool, impounded on one of the state's few reservations, or, if they proved not docile enough and demanded too much in the way of human dignity, exterminated.[65] And though the specifics were unique to California, the Golden State by no means cornered the market in treacherous dealings with American Indians: broken promises, atrocity, and ethnic cleansing stretched back centuries. The bloodiest war in American history, King Philip's War, was fought in colonial Massachusetts in the mid-1670s, and it was a war that helped frame all future imperial conflicts, a war that helped cement the view of American Indians as inherently inferior to whites.[66]

This was an ugly and all-too-public history, one few wanted to face. Much better if American Indians could be turned into supporting play-actors in the American drama of Manifest Destiny. Much better if, like Bunnell's Ahwahneechee, Native people were a vanishing race of noble savages, compliantly melting away like the dawn's mist before American enlightenment. It was not war that was exterminating the American Indians, according to the myth, but divine destiny. And yet, the very best characteristics of the savages, their stoic, individualist, masculine nobility, could be appropriated by white Americans looking to critique the perceived effeminacy of a decadent Europe: playing Indian, from the costumed agitators of the 1773 Boston Tea Party to the chanting and tomahawk chopping of today's Washington Redskins football team, was, and is, a way to perform American uniqueness.[67]

Both of these narratives—the vanishing race and the noble savage—are at work in the California explorer and geologist J. D. Whitney's 1868 elegiac pronouncement that "[Sequoyah's] remarkable alphabet is still in use, although destined to pass away with his nation: but not into oblivion, for his name attached to one of the grandest and most impressive productions of the vegetable kingdom will forever keep his memory green."[68] The noble savage and the vanishing race, in other words, are ways to wish away the brutality of murder and dispossession, and we could see both tropes grafted onto another foundational American narrative—call it the myth of the ecological Indian, a story that sees native peoples as closer to the natural world, as less evolved than Westerners.[69] Naturalizing race has always carried with it the connotation that nonwhites are slightly less than humans, a bit lower on the great chain of development, closer, perhaps, to giant trees than to Euro-Americans.[70] Asa Gray could have been writing about American Indians when he asked, "are [the sequoias] veritable Melchizedecs," comparing the trees to an ancient Jewish priest who was neither born nor would ever die, but would live in a preternatural suspension, forever, "without pedigree or early relationships, and possibly fated to be without descent. . . . Or are they remnants, sole and scanty survivors of a race that has played a grander part in the past, but is now verging to extinction?"[71]

Yet, all three storylines of American Indian as vanishing, noble, and ecological are narratives of forgetting and erasure; none explain the howling irony of why the Big Trees became a monument celebrating a people in the midst of being systematically exterminated. Here's the paradox: midcentury America, "Nature's Nation," had staked its identity on both nature and on a sense of Indianness; and yet both forests and American Indians were falling to the axes of Manifest Destiny, leaving national identity orphaned.[72]

One way to resolve this paradox is through what the historian of theater Joseph Roach has called "surrogation." The cultural history of the New World, Roach argues, is indelibly marked by constant loss—American Indians are removed, slaves are uprooted, places are bulldozed under. How is it, then, that culture has flourished, especially among those most affected by loss? To answer that question, Roach turns to the theater and asks what happens when the lead actor in a play falls ill. Some one—an understudy—fills in and takes her place, assuming multiple roles. And the show goes on, as it must—this is how surrogation works.[73] If American national identity relied on both American Indians and forests, then one way to preserve both at the same time, even as both were being obliterated, was to collapse them into each other, to make one a surrogate for the other—a wooden Indian, a tree named Sequoia. Even the word *savage* plays this shape-shifting game: it originally derives from a cluster of words meaning "forest," or "tree."[74] Ever since the colonial era, trees and American Indians had seemed interchangeable to many Euro-Americans: "The trees stood Centinels and bullets flew / From every bush (a shelter for their crew). . . . Every stump shot like a musketeer, / And bows with arrows every tree did bear," wrote the Boston poet Benjamin Tompson of King Philip's War.[75]

And so it was a complicated play with four plot lines that all boomeranged between forgetting and remembering that was staged beneath Big Tree boughs: American Indians as vanishing, as noble, as natural, and, finally, as sylvan. As long as there were a few ancient trees left standing to perform the role of American Indian, there would also be a talismanic Native presence; and because the very attributes that made the sequoias seem like such a natural symbol of the new nation were also many of the same characteristics idealized in the image of the noble savage—their great antiquity, their durability, their strength, their endurance—the sequoias were to remind Americans of their noble ancestry, even as indigenous people slowly exited the scene.

Such sequoian witnesses to an American scene took on added urgency in the post–Civil War United States. During the conflict, Free Soilers drafted sequoias into the Union's ranks, trees with names like Uncle Tom's Cabin, Abraham Lincoln, U. S. Grant, Henry Ward Beecher, and General Scott, an American Birnam Wood waging cultural war against the South.[76] But when the guns finally fell silent, and Americans looked over their shattered landscape wondering how to heal, they again turned their eyes treeward.[77] In a country recently rent by war, Sequoyah, the accommodationist, asked white citizens of the North and South to bury the hatchet along with their dead, to turn to a destiny long foreseen by nature and American Indian, to stitch the United States back together, to get on with the divine work of America: seizing territory and making money. And so after the war's end, the very trees themselves were made to proclaim national unity: alongside U. S. Grant and Abraham Lincoln stood General Lee and his group of five Confederates, their sylvan reunion symbolizing one unbroken spatial, temporal, sylvan Manifest Destiny with the sequoias as the indisputable culmination of the type, "the king of all the conifers in the world, 'the noblest of the race,'" as John Muir put it.[78]

6

The sequoia's tale of irony isn't quite done yet, for in 1879 the largest of the world's biggest trees was baptized General Sherman, named for General William Tecumseh Sherman, a man whose legal first name, until he was nine, was simply *Tecumseh* after the Shawnee leader who tried to forge a pan-Indian alliance across the Mississippi Valley in the early nineteenth century to resist white incursion; Tecumseh Sherman found fame as the Civil War general whose scorched-earth policy of total war left his route from Atlanta to the sea a smoldering wasteland— though in his own day he was perhaps only slightly less famous as the post–Civil War general whose seething hatred of the American Indians seeped into his private letters, where he wrote that the Indian wars were "one of those irrepressible conflicts that will end only in one way, when one or the other must be exterminated."[79] "Treachery is inherent in the Indian character," he told a *New York Times* reporter in 1873, reacting to news of the war with the Modoc Indians of northern California. Three days earlier he had telegraphed his commander in charge of field operations against the Modoc: "Make the attack so strong and persistent that their fate may be commensurate with their crime. You will be fully justified in their utter extermination."[80]

Extermination is a word Tecumseh Sherman never used in reference to his Confederate enemies. It crops up with a grisly frequency whenever he writes of American Indians.

7

Trees are unreliable narrators. They bend under great weight, and, pushed too far, they're likely to snap. A tree can't ever be an American Indian, and an Indian killer can't ever be a sequoia. Some surrogates, some replacements, are simply too grotesque: rather than conceal, they remind us of what has been lost.[81] And so even as so many labored so diligently to turn American Indians into mere sylvan traces, trees called sequoias refused to fully obscure the indelible blot of imperialism; they stood like silent monuments, and one didn't have to linger long among them before feeling the brutal absurdity of a sequoia named Sherman.[82] Tales of haunting have long taken place in a forest.

For all the blithe pronouncements on the divine white Americanness of the sequoias, there was a nagging notion that sequoian frames couldn't contain the burden of the past. One feels it explicitly in the most frequent response that tourists had when encountering the Big Trees, an anxiety about what secrets they held, what testimony they would tell. When Thomas Starr King, who had a tree named after him, wondered, "Are you as old as Noah? Do you span the centuries as far as Moses? Can you remember the time of Solomon?" there's a quaver in his voice, a fear that the past's secrets might be no more secure than the bundle of letters in the old shoe box.[83] It's a worry that Hutchings couldn't quite shake, either, even with an exclamation point: "Could those magnificent and venerable forest giants of Calaveras county be gifted with a descriptive historical tongue, how their recital would startle us, as they told of the many wonderful changes that have taken place in California within the last three thousand years!"[84]

8

What is a dream but the feeling of the past or the future, some alternate reality intruding upon the present, a feeling of being haunted?

Dreaming makes rational time unhinge, causing the this-then-that arrow-straight linearity that characterizes much history to convolute in squirming coils; points on the timeline, rather than remaining distant, double back upon each other and into near contemporaneity. When we dream, we connect in defiance of chronology.

A dream is the hallucinatory space of anachronism into which possibility rushes.

Writing history is a kind of lucid dreaming, and all histories are masterpieces of snatched moments stitched together.

Dreaming is the antithesis of irony.

Dreaming is full of potential.[85]

9

It was nearly inevitable that alternative socialist dreams would take root among the sequoias, for the specter of socialism haunted the star of capitalist empire as it shot its way west: after all, the very first attempt at founding a socialist colony in the United States was made by the Dutch Mennonite Pieter Cornelius Plockhoy in 1663 near present-day Lewes, Delaware; and besides the communes nurturing radical abolitionists and freedom-loving anarchists in the midcentury East Coast, the Owenites, the Associationists, and the Fourierists—socialists all— gradually extended their hands ever-westward.[86] Greeley, Colorado—

now infamous for the stench of its stadium-sized confined animal feeding operations and the dehumanized, exploited workers who keep cheap meat on America's dinner table—was founded as a utopian community by Horace Greeley, the most influential journalist of the mid-nineteenth century. "Go west, young man, and grow up with the country," we remember Greeley for saying, and many socialists followed the advice: between 1800 and 1914, there were something like 260 viable secular socialist societies throughout the United States, and if we count Catholic and Protestant religious orders, Shaker and Mormon communities, Rappites, Icarians, and the dozens of small socialist ethnic religious enclaves, there is a good argument to be made that communalism, as much as rugged individualism, has been the key to American development, and that socialist communes are as American as a protest march on Washington.[87]

Though the widespread popularity of communes waned somewhat in the aftermath of the Civil War, interest in alternatives to economic individualism and mainstream politics, especially after the economic Panic of 1873 and the great railroad strikes of 1877, the most violent labor upheaval up to that point, continued to grow. The California transplant, Henry George, that critic of the transcontinental railroad (whose 1879 work *Progress and Poverty* argued, in an almost ecological way, that exploitation was not part of an ethical life, that land was the real and only source of wealth and thus everyone had an inalienable right to it) was widely read and spurred intense leftist political discussion throughout the country.[88] On the cultural front, Edward Bellamy's 1888 socialist novel, *Looking Backward*, was a best seller despite its wooden narrative, and Bellamy Clubs, 162 in all, sprang up throughout the country, urging a nationalization of all industry.[89] In California alone, there were at least nine utopian socialist communes at work between 1850 and 1900.[90]

And so the empty, parklike landscape surrounding Hale Tharp's former hollow-log home, with its Giant Forest and General Sherman came to be home for an anticapitalist band of radicals yearning for a free breath of clean air. "The scenery is sublime," boasted the anarchist and socialist Burnette Haskell when he first saw the land in the mid-1880s; "the eternal white-capped mountains of the Sierra Nevada are in full view, while the mountain gorges and canons [sic], with their majestic water-falls, are spoken of in raptures by those who have been fortunate enough to view them."[91]

Haskell's comrade, J. J. Martin, caught the message of change carried on the breezes tumbling from the summit of the Sierras, relayed through the sequoias, and into his ear. It was a beautiful spot, so different from his native England and the other American cities in which he had lived: Galveston, New Orleans, San Francisco.[92] It looked and felt to the Englishman like the sort of place from which a glorious future could be nurtured, the new world's fresh green breast sustaining eutopian cooperation. "The heat is greatly modified by refreshing breezes," he wrote to his fellow socialists: "The nights are perfectly glorious. The air is pure and invigorating. Springs of cool water are plentiful, and shade is abundant."[93]

Haskell, Martin, and their comrades had come to a place called Kaweah, a name that some thought referred to the Gaiwa Indians, a band of Yokuts who lived along its banks; or it may have meant "raven" in Yokut, or perhaps the "river of the calling raven," or, as Haskell liked to tell it, "here we rest."[94] Whatever the true meaning, the Kaweah colonists felt that the nobility of the American Indians for whom Kaweah was named could be regained, could be used to scrub clean the filth of competitive individualism into which the race for profit had plunged the United States.[95] It was a new beginning.

Kaweahan dreams had been hatched back in 1884, at a picnic hosted by the San Francisco chapter of the Knights of Labor.[96] It was here that Martin discussed the labor question and its possible solutions with his friend the organizer, lawyer, and newspaperman Haskell, and about a hundred others. Haskell had been a devotee of the anarchists Pierre-Joseph Proudhon and Mikhail Bakunin, as well as of Karl Marx, and in 1882 he founded a secret socialist reading group named the Invisible Republic; at around the same time he also began printing his newspaper, *Truth*. By 1883, his radicalism aglow, Haskell created a revolutionary organization based loosely on the Marxist First International called the International Workmen's Association, an organization that sought to bridge the rift between anarchists and socialists, and that would in time become fairly important in California labor struggles.[97]

A revolution without a picnic is not a revolution worth having, and at this one ideas for replacing capitalism were passed back and forth along with the wine and bread, each a different, spicy dish: some favored education, others immediate and violent struggle, while still others felt that letting evolution take its course was the best path to the good life. Underneath all the heated rhetoric, the posturing, the competition to be the most radical—which I am sure must have been on display (it depressingly comes out in the later records)—there is a profound philanthropy at work. Looking back on the birth of Kaweah, one of the colonists would write, "If faith in humanity is lost, *all is lost*, and there is no hope left."[98] A few months after the picnic, the very beginnings of a plan were put in motion: an organization would be formed, called the Co-Operative Land Purchase and Colonization Association, which empowered each one of its members to "consider himself a committee of one to seek out opportunities to purchase."[99]

The idea was to create a space completely different from the capitalistic, competitive arena of the gladiatorial free market, one that integrated work, leisure, and living, and which allowed for the fullest development of each individual even as it fulfilled the social needs of security, health, and industry. Using Laurence Gronlund's newly published *The Co-Operative Commonwealth: An Exposition of Modern Socialism* (1884) as their bible—a work that was the first distillation and adaptation of Marx's *Kapital* for an American audience—Martin, Haskell, and their comrades launched their crusade.[100]

And so the revolutionaries fanned out, eyes and ears alert, searching for a plot of land from which they could stage their peaceful revolution. Thanks to disease, violence, and the dislocating pressures of the gold rush, the California landscape seemed a fallow wilderness: by 1880, California Indian populations numbered only around twenty-three thousand, down from about three hundred thousand when Spain began colonizing the area a little more than one century before.[101] Land that had previously belonged to American Indians, the Spanish, and Mexicans, if it hadn't already been appropriated by individual owners or the railroads, was owned by the federal government, which sought to turn those perfectly square sections into taxpaying farms through the passage of various bills—but there was a catch.[102] All of this legislation sprang from the utopian Jeffersonian faith in the small, yeoman farmer—a mythical creature in a late nineteenth century that saw the rise of truly massive corporate factory farms—and so care was taken to ensure that the land went to individuals, rather than corporate entities; federal land agents, at least the honest ones, were constantly looking for anything that smelled like an attempt at land monopoly.

In 1885, C. F. Keller, a member of the Co-Operative Association, was steaming back to San Francisco from a land-scouting trip on a Southern Pacific train, and as luck would have it, he happened to overhear two surveyors—or speculators?—discussing land in and around the Giant Forest, directly in General Sherman's shadow.[103] Timber cruisers for the bigger logging operations had declared the land inaccessible, and therefore no one had though to snatch it up.[104] It simply wouldn't pay. But for socialists, the prospects were promising, and when Keller reported his find to Martin and Haskell, they reacted quickly. On October 5, 1885, forty-two radicals marched to the land office in Visalia, California, and filed their claims for a stake in and around the Giant Forest. With something like twelve square miles of land to their collective names, and with a plan to merge all of the land under one entity, the Kaweah Colony was no longer a dream but a reality complete with demarcated boundaries.[105]

To get to their colony, the socialists rumbled east from Visalia, the seat of Tulare County located in the fertile San Joaquin Valley, and out along the rough road to Mineral King, a boomtown built by those hoping to blast their way to wealth. The road clambered up the rungs of the Sierras, and after twenty-five miles, came to the small foothills village of Three Rivers, where the North, Middle, and South forks of the Kaweah river diverged. Had they continued along Mineral King Road, they would have passed a sequoia called Witness Tree, so named in 1882 by the men who had been working on the road; they carved their names into the tree's trunk.[106] But they didn't; they instead took a left, hugging the north fork of the river until they alighted at the spot that would soon become the Kaweah Townsite. When they did, they looked around in awe at "a place of great spring beauty and always of charm," as Philip Winser, an English hop picker who spent his nights pouring over Thoreau, Emerson, and Louisa May Alcott, remembered. Winser had left his native England to join the colony in the later 1880s, and he wrote that the landscape seemed to cradle the infant revolution: "The hills shut us in on all sides ... and the sides were flanked by chamiso, chapparel and oak dotted hills, with manzanita and buckeye fairly plentiful. The former carried pretty pink and white clusters of heathery bell like bloom in spring, having a honey fragrance."[107] The area had been used to graze sheep and cattle since the 1850s, and it was renowned for its richness, with water, sunshine, cool weather, marble deposits, and, for the scenically inclined, Moro Rock, a huge granite dome in the center of the colony, which provided unparalleled views of the Sierra and the sequoias.[108] Most important, there was timber in great supply. Everything was at hand for the communards' "step higher in the scale of human existence ... the dawning of the long prophesied millennium."[109]

Dreams are often promiscuous, and in the Kaweahans', art and science grow in forests; nature's evolution gives rise to their socialist political economy; land and labor ensure each other's integrity—the colonists were tantalized by the possibilities of consorting with nature. And so when the Kaweahans took possession of their land, one of their first acts was to stream into the Giant Forest and exorcise the personification of murder and racism from the largest tree in the world. In place of the bloody General Sherman, they substituted a new name: Karl Marx.

Rather than the eutopian agrarianism of Timbuctoo, the Kaweahans understood that Marx's sequoian comrades themselves were prophets of a breaking future, and the trees under whose branches they lived were a crucial part of Kaweah's social vision. The Kaweahans would log among the Big Trees.

This might seem sacrilegious to today's environmentalists—Sierra Clubbers and tree sitters, alike. Why not, while we are at it, hammer away at a cathedral for paving stones, or sow our fields with salt? But a leave-no-trace environmentalism, however well it may be suited to

FIGURE 41. The Kaweahans and Karl Marx. Photo by C. C. Curtis, ca. 1887. Yale Collection of Western Americana, Beinecke Rare Book and Manuscript Library.

backcountry wilderness trips, has almost no relevance to the social, cultural, and environmental problems that we all find around our cities, farm fields, roadways, factories, and dumps. We must leave a trace if we are to live, and we need an environmental ethics that can help us to leave our marks, healthfully and justly.[110]

This is what makes the Kaweah Colony so radical: they proposed an economic, social, and environmental system that could sustain human as well as nonhuman life; their vision of landscape was not simply a means of walling nature off from humanity, but a method of interconnection. Whereas the other logging camps in the Sierras were just that—camps, temporary squatting grounds for exploited migrant laborers run by owners living somewhere comfortably removed from the work of stripping the land—Kaweah was a home: the communards would be logging their own backyard. And while trees would have to be cut, the Kaweahans also understood that raw materialism, "cold blooded calculation" as they put it, was not enough to nurture a good life; ancient trees were more than simply so many board feet of lumber—they were mentors—and the communards frequently wrote that they had no intention of felling Karl Marx or any of his sequoian comrades: "It would be nothing short of vandalism to indiscriminately destroy these sentinels of past centuries, as has been done in several parts of California, by ruthless ravagers of the Competitive system and care will be taken to preserve them in their primitive glory."[111] Indeed, their sawmill was located nearly eight miles from the biggest stands of trees, though their road did continue into the Giant Forest itself.[112] "None of the big trees were to be touched," wrote J. J. Martin: "They were regarded because of their age and size as sacred."[113] Instead, they fought fires, and even proposed to build a scenic hiking trail so that outsiders could admire the landscape of socialist cooperation. Besides, sequoias make poor lumber: the Kaweahans knew this, even if many contemporary California boosters chose to ignore this fact in the pursuit of a hot dollar.[114] Not only were the biggest of them far too large for the mills of the day—the other logging operation in the Sierras blew the trees apart with gunpowder—the wood was brittle, and not very desirable for construction. Much more suitable were the stands of sugar pine, spruce, and fir. The economist W. Carey Jones, with a bit of the economists' characteristic hyperbole, noted that the Kaweahans' land was thickly forested enough with these other more suitable species of trees to supply "an immense market for many years."[115]

"The labor movement is profoundly impressed with the spirit of the age," wrote Haskell after the colonists had been harried from their land at bayonet's point. "Whether it know it or not, the bugle of evolution has given the guide to every file of its broadening and marching flank... and we, its dreamers, were but a skirmish line of the main body."[116] If there's a sense of historical inevitability threading its way through the socialists' thinking, it's also true that this thread reinforced the connection with the outside world, and one way to see the Kaweahan project is as an example of regional development: the first of two ambitious projects began almost immediately when the colonists conceived a plan to build a railroad from the Giant Forest down to the Southern Pacific's mainline, a way for socialist lumber to infiltrate the mainstream of American commerce. Without the massive government support that the robber barons enjoyed, a socialist iron horse proved impractical in the short term, but in 1886 the colonists began building an eighteen-mile wagon road from Kaweah Townsite up into the stands of timber.[117] Farmers and settlers in the San Joaquin Valley may have had perfect weather and deep soils, but they weren't blessed with an abundance of trees, and the Southern Pacific Railroad was charging a mint to ship lumber from the northern part of the state into the valley, despite the fact that plenty of timber stood only a few miles away, in the foothills surrounding Kaweah. "There is not a building of the better class constructed entirely of material produced in the county," wrote the editor of the *Historical Atlas Map of Fresno County*, in 1891. "In a frame house it will be found that the rafters, studding and floors came from Puget Sound, the wainscoting, rustic, and shingles, from Mendocino, the doors and sash from Shasta," all profitably shipped along the Southern Pacific's tentacle-like limbs.[118]

At the same time, the Kaweahans launched their second project: putting into practice a political economy that, they hoped, would grow from the Sierras and eventually, by dint of its practicality and fairness, spread throughout the nation. This was the time-check system, a way of fundamentally reorganizing work—and thus one's relationship to the nonhuman world—along cooperative lines, and it came swaddled in all sorts of revolutionary sayings intended for the picket sign, like "cost is the limit of price."[119] The concept was simple. All wealth, the Kaweahans thought, flowed from two sources—land and labor—and money was but a poor approximation of value. Instead, Kaweah's economy would revolve around the time check, a system whereby the time and "life force" expended in any job, including the artistic and intellectual, would be fully repaid the laborer, with no margin of profit, thus eliminating the temptation to exploit bodies or natural resources. Without the profit motive to goad a person's greed, the things the Kaweahans would sell to the outside world—like lumber—would be cheaper. Nor would it be possible to buy low and sell high, because the value of any product would not fluctuate. The more time and the more sacrifice it took to create a good, the more it was worth, and those who worked the longest at the hardest tasks were rewarded the most richly.[120] It was a eutopian scheme, and today, it's easy to scoff about the practicality of scientifically defining "life force" (is the dollar a better measure?), but even the hard-boiled economist Jones wrote, "On the whole, I can say, although the comparison does not do them justice, that their life is fuller, better, and more profitable than that of either the average California farmer or the members of the average California village," and it proved popular.[121] J. J. Martin reported that Kaweah, at the time of its dissolution, had 615 members—though never more than 150, and more frequently somewhere between 50 and 75, actually lived in the forest.[122]

It might seem like the Kaweahans spent the majority of their intellectual steam rethinking labor and compensation, and that they left the other half of the basis of wealth—land—behind. Yet underlying the Kaweahans' time check was a mixture of radical thinking derived from, among others, Pierre-Joseph Proudhon's contention that private property is thievery and Laurence Gronlund's argument that all products "have sprung from nature, and contain in them a certain amount of human labour."[123] When Proudhon denounced private property, what he meant was that the productive capabilities of the land, because they inhered in the earth itself, could belong to no one. No one could own the earth, no one could privatize the land, turning it into property, though Proudhon conceded that one could certainly own his own clothes or her own tools or books. But the earth and its fertility was given equally to everyone. The process of seizing land, wrote Proudhon in his *What Is Property?* (1840), renting it out to others to work, and then claiming a portion of the workers' harvest as payment was completely immoral, on three levels: since everyone had an equal claim on the productive capabilities of land, seizing it and declaring it one's own was a first case of theft. To then rent the land back to those from whom it had been taken was a basely cynical second theft. Finally, to claim from the laborer a portion of that which was truly his—that thing he had made or grown through his own genius and sweat in combination with the land's fertility—under the pretense that the landowner had some natural right to whatever sprang from the earth, was antisocial theft, pure and simple: "Property and society are utterly irreconcilable institutions," wrote the French anarchist. "Either society must perish, or it must destroy property."[124]

A similar reasoning runs through Gronlund's *The Co-Operative Com-*

monwealth, and this foundational intellectual plank—that property is theft—is the radical environmental principle at work in Kaweahan thinking, one that refused to write environmental considerations off as mere externalities to the more important working of the market. Because land belongs to all of us, Proudhon wrote that the laborer "is responsible for the thing entrusted to him; he must use it in conformity with general utility, with a view to its preservation and development; he has no power to transform it, to diminish it, or to change its nature. . . . In a word, the usufructuary is under the supervision of society, submitted to the condition of labor and the law of equality."[125] This is a sort of mutual aid, an idea of evolution guided by cooperation that predates both Darwin and Kropotkin, and it is the sort of environmentalism that was radical because it was so expansive. One might object that it's also an unreconstructed anthropocentrism, which is true enough. But it's also important to note that Proudhon refers to his ideal laborer as a usufructuary, as someone who has a moral duty to leave the landscape as productive, as healthy, as beautiful as she found it precisely because the land must always remain unowned. The productive capabilities of a stream, a field, or a forest must never be diminished. For the anarchist Proudhon the greatest guarantor of an individual's right to grow unfettered according to her own genius lay in community and in ensuring that the wealth of the world was never swallowed by a rapacious few, but remained available, to everyone, forever—a sort of ultraradical sustainability. In this sense it anticipates Aldo Leopold's famous, though far more circumscribed, land ethic: "A thing is right when it tends to preserve the integrity, stability, and beauty of the biotic community. It is wrong when it tends otherwise."[126]

This is what made Proudhon and Gronlund and their Kaweahan followers such radical thinkers: enclosing the land and seizing the products of another's labor is wrong because, in the end, it impoverishes the integrity, stability, and beauty of the entire biotic community—the individual laborer, the trees, the animals, the landscape, and society, all. Unlike today's mainstream modern environmentalism, which limits its concern solely to the nonhuman world of soil, trees, animals, and climate, the Kaweahan focus on the points of human-nature interaction and mutual aid foreshadows the green tint of social ecology. "The relations of the State," wrote Gronlund, "to its citizens, is *actually* that of a tree to its cells."[127] Everything that lives has a desire to keep living, and to do so healthfully, according to its own genius; we all have a responsibility to both society and the land. The alternative, as Gronlund put it, was a mere instrumentality, a system that dismembered both humans and nature: "The labouring men are dealt with by our managers as mere tools. They are spoken of as tools, as things. This humanitarian age counts steers and sheep by 'heads' and the workers by '*hands.*'"[128] If John Muir's brand of environmentalism tells us what not to do when we are in the woods, and is silent when we return to the cities, Proudhon and Gronlund follow us into the forests, through the fields, and into our backyards, into our very social arrangements, the very sorts of places Muir and even Aldo Leopold feared to tread.[129]

10

Kaweah. The gently drawled, cool evening breeze of a vowel cluster, is rarely uttered anymore, except by a few specialists in nineteenth-century US history and a small handful of back-to-the-landers with long memories. Forgetting was the result of an unlikely combination of interests working to undercut Kaweah.

First, newspaper editor George Stewart, who would become known as the "father" of Sequoia National Park, began heatedly advocating for a nature reservation in the Giant Forest.[130] Ever since 1878, he had been lobbying anyone who would listen for the preservation of the Big Trees, and he was bolstered by Abraham Lincoln's 1864 decision to grant Yosemite Valley and the Mariposa Grove of Big Trees to the state as a park. By the 1880s Stewart's vision had come to encompass a Sierran alpine preserve consisting of tens of thousand of acres, and many farming interests in the San Joaquin Valley actively supported Stewart's crusade because they had been persuaded, rightly so, by a flood of environmental thinking reaching back to George Perkins Marsh's epochal *Man and Nature* (1864), that without forests, the mountain water irrigating their fields would dry up before it had the chance to percolate into the ground.[131] John Muir, of course, joined in on the crusade for preservation, and likewise latched on to the benefits a park would render to farmers: "The value in these forests in storing and dispensing the bounty of the mountain clouds is infinitely greater than lumber or sheep," he wrote. "To the dwellers of the plain, dependent on irrigation, the big tree, is a tree of life."[132]

Punctuating these voices calling for a park was an unsettling bureaucratic throat clearing: it turned out that the federal government had never granted ownership over any of the sequoian land to any of the communards. Since 1885, they had been squatting.

The problem was that on that October 1885 day when forty-two eager socialists filled the small land office in Visalia with their hopes and their land claims, it raised suspicions that these were not earnest settlers, but dummies purchasing land for a giant capitalistic timber corporation.[133] It was the soon-to-be father of Sequoia National Park himself who first raised the alarm. These weren't idle fears: every land act, from the Preemption Act of 1841 to the Timber and Stone Act of 1878 saw massive fraud; indeed, corporate timber interests—including those directly related to the Southern Pacific—had already snatched up much of California's forests by such crooked methods. Xenophobia also played a role, as it often has in environmental politics: some of the Kaweahans were either not yet US citizens, or simply had too-foreign sounding names, and Stewart would not believe that such people could come up with the filing fee for their land unless they were stooges for big timber.[134] And so he alerted the authorities.

When the land commissioner in Washington, DC, was apprised of the situation, he suspended the land claims, pending further investigation, which, in a long and convoluted process, first determined that the Kaweahans were entirely legitimate, and then that they entirely weren't.[135] Nonetheless, many locals actively supported the socialists; in fact, it was the official land office agent in Visalia who advised the communards to squat until everything was cleared up, as surely it soon would be. If they improved their holdings in the meantime, all the better, for it would go to show that the Kaweahans had filed their claims in good faith and intended to stick.[136]

It turns out that the Kaweahans' fatal error was that they weren't as paranoid of their capitalist foe as they should have been. In 1885 or 1886, when the Kaweahans were dreaming of building a socialist railroad from the Giant Forest to the Southern Pacific's mainline, J. J. Martin strode into the office of the railroad's president, Charles Crocker, to ask for help. Crocker had been one of the big four investing in the Central Pacific Railroad, the western half of the transcontinental railroad, and had realized unimaginable wealth from his early investments. He was one of the most powerful men in the country, running one of the most hated monopolies in California, and it took either an appalling lack of forethought or an incredible faith in one's own mission for Martin to march into the sumptuous office of one of the country's most successful and ruthless capitalists, announce one's socialist, antibusiness sympathies, lay out a plan by which capitalism would eventually be overthrown, and then propose a collaborative business venture that would surely cut into the capitalist's earnings; it took either guts or naïveté to do all this and expect none other than Charles Crocker to lend a helping hand—but that's exactly what Martin did.[137]

Did Crocker smile? Was there a nervous flicker in his eye?

He certainly took Martin seriously.

The local Southern Pacific land agent for the San Joaquin Valley and the foothills of the Sierras—Kaweah country—was Daniel K. Zumwalt, one of Southern Pacific's fixers. Zumwalt actually lived in the same county as the Kaweahans. By the 1870s, he had been instructed to personally investigate all new land claims in his bailiwick—these claims were personal challenges to Crocker's authority—and he had been on the scene in 1880 when the railroad's men gunned down five farmers at Mussel Slough in a dispute over land ownership.[138] Zumwalt was also a preservationist and conservationist in favor of preserving the Big Trees; as early as 1889 he had signed his name to a petition urging the state to withdraw an area of the sequoias from the land market so that the San Joaquin Valley's farmers might not see a reduction in their water, which would mean fewer crops, and thus less cargo for Zumwalt's employer. Zumwalt, then, most likely knew the Kaweahans' plans long before Martin ever set foot in Crocker's lair.[139]

Then, in October 1889, at the request of Southern Pacific, the Californian congressman William Vandevar introduced a bill to nationalize Yosemite and the Mariposa Grove of Big Trees state parks. But the bill languished in Congress, despite intense lobbying by the Southern Pacific, which was hoping to capitalize on the increased tourism a national park would bring.

George Stewart seized this very moment to present Vandevar his own bill, one proposing Stewart's longed-for Sequoia National Park.

It was the right moment to strike. The new Sequoia Park bill raced through Congress, and was signed into law, without debate or amendment, on September 25, 1890.

Back in California, the Kaweahans greeted the Sequoia bill with joy: the lands reserved were just to the south of their holdings and preserved some of the finer trees in the Sierras, forever. It seemed that the federal government was joining the revolution.

However, only a week later, on October 1, 1890, another bill was quietly signed into law. It was the Southern Pacific's earlier, stalled Yosemite National Park bill, but with a significant new addition: a rider tacked on a huge expansion to the week-old Sequoia National Park, one that included nearly all of the Kaweahans' land.

No one really knows how or why the Sequoia National Park enlargement made it into this Yosemite bill, but almost everyone agrees that the Southern Pacific was behind it.[140] Not only had the company sponsored the original Yosemite bill, but Zumwalt was in Washington during the bill's passage, peddling influence and staying with his good friend—Congressman Vandevar.[141] Richard Orsi, Southern Pacific's hagiographer, writes that "Zumwalt . . . had the bill amended at the last minute to add provisions more than doubling the size of the giant-sequoia park," but as it turns out, almost no one knew that this language was actually in the bill—except for Zumwalt and the railroad.[142] The bill referred to the proposed reservation only by township number, and when Senator George Edmunds of Vermont asked for the bill to be printed so that he could study it more clearly and figure out what he and his colleagues were really voting on, someone, we will never know who, took him aside, after which Edwards withdrew his request.[143] Once the bill was passed, no one knew what the implications were—except for Zumwalt and the Southern Pacific, which quickly had a map of the enlarged park and its proximity to their railroad lines drawn up, long before anyone in California knew what had happened.[144] In the cruelest twist of the story, on November 24, 1890, two days before Thanksgiving and six weeks after their dream had been stolen, the local Visalia Bank had agreed to fund a local railroad up to Kaweah. "Work will be commenced next April," wrote Haskell's wife, Anne, elatedly. "It will be a great thing for the Colony."[145] At nearly the same time, photographers and painters started popping up in the colony, courtesy of the Southern Pacific, not to photograph the achievements of socialism as the Kaweahans thought, but to create advertisements for the new park.[146]

Though there is no smoking gun, the motive behind the enlarged Sequoia National Park was not preservation at all—or rather, it *was*: preservation of the Southern Pacific's wealth; preservation of capitalism.

Gold and sequoia.

The sensible thing for the Kaweahans to do would have been to concede defeat; but the thing that I have come to learn and admire about eutopians is that no matter how loud the capitalists trumpet their triumph, eutopian dreams for tomorrow's better world can't be killed. The colony, unwilling to heed their own funeral dirge, actually struggled on until April 1892, and for a time, the Kaweahans continued to cut timber on the land they claimed was theirs, all the while sending petitions to Washington. But then in December, four communards were arrested for timber rustling and shipped off to Los Angeles to stand trial.[147] When, later that year, Kaweahans once again resumed cutting wood—this time on a legally patented claim some twenty miles distant from their disputed lands—it took fifty-eight members of the US Calvary to harass the colonists, who, the officials back in Washington agreed, were doing nothing illegal, into nearly giving up.[148] It wasn't until late 1892, when the trustees of the colony were indicted for mail fraud—they had been sending requests for donations through the post office, a heinous crime for radicals — that the eutopians left their forest home, leaving the ground for Sherman's men who finally expelled Karl Marx, reinstalling General Sherman in his place.[149]

Incredibly, the Kaweahans were still not beaten, not unconditionally. For a while, many considered relocating to Mexico, and J. J. Martin even left to start communes in British Columbia and Tasmania. Martin spent the rest of his life lobbying the government to clear the Kaweahans' name and to reimburse them for the illegal seizure of their land.[150] Kaweah inspired other Californian communes, too: Winter's Island and the Army of Industry both took encouragement from Kaweah's example, and even today, at Virginia's sylvanly named Twin Oaks, one of the longest-running communes in the United States, there's a building named Kaweah.[151]

11

Every funeral needs music, and for Kaweah's, Allen Dodworth's "Woodman, Spare That Tree!" fits. In 1848 (that hopeful year when Marx and Engels wrote "A spectre is haunting Europe — the spectre of Communism," and Thoreau wrote "Civil Disobedience," and Gerrit Smith reported that twenty or thirty black families had settled in the Adirondacks), Dodworth borrowed George Pope Morris's popular 1837 poem of the same name and set it to music, a snappy up-tempo ditty played in the left-right-left marching time signature of 2/4 set at an *allegro marziale*, a lively martial pace.[152]

Though it is of course unfair to judge sheet music by its cover, the illustration accompanies Dodworth's tune well: a scene set beyond the fields, in the wild woodlands, literally at the end of the road, where a city gentleman, the character we are meant to identify with against the rude country rustic whom he nearly tackles, arrests the blow of the woodman's ax. March, Dodworth tells us. Hurry up and with a stiff spine defend the woods from the ignorant.

FIGURE 42. Cover lithograph for Allen Dodworth, *Woodman Spare That Tree! Quick Step, as Played by the Dodworth Coronet Band* (New York: Firth, Pond & Co., 1848).

This bit of sheet music is, I think, the perfect soundtrack accompanying the rise of mainstream environmentalism towards the end of the nineteenth century, the sort of tune to which George Stewart, John Muir, Daniel K. Zumwalt, Charles Crocker, and the US Cavalry—which guarded both Yosemite and Sequoia National Parks for twenty-five years before the National Park Service was created in 1916—would have thrilled.[153] The early history of mainstream environmentalism is filled with elite condescension aimed at those living in rural areas, a separation of the world of work from that of play, the country from the city, an antidemocratic sentiment that certain enlightened individuals always know best, and that the human hand, the working hand, the hand as metonym for a laborer, is only ever destructive.[154]

And so perhaps here, at the very end, the sequoia's bitter irony returns. Yet Morris's original poem is far subtler than Dodworth's tune suggests:

> Woodman, spare that tree!
> Touch not a single bough!
> In youth it sheltered me,
> And I'll protect it now.
> 'Twas my forefather's hand
> That placed it near this cot;
> There, woodman, let it stand,
> Thy axe shall harm it not.

There's a sense of intergenerational, interspecies community and responsibility in Morris's telling, and though it seems that the poet has moved from his ancestral cot, his sense of place hasn't changed. The "hand" is a husbandman's, a caring hand, a hand that works in harmony with the landscape to rear and protect something nonhuman but nevertheless familiar. The tree is a third parent to the narrator and his sisters, and is the only witness to a nearly-forgotten family affection: "My mother kissed me here; / My father pressed my hand." The past is gone, but the sacred tree yet lives, which, for the poet, means that the past has been resurrected, and he is haunted by reciprocal obligation:

My heart-strings round thee cling,
Close as thy bark, old friend!
Here shall the wild-bird sing,
And still thy branches bend.
Old tree! the storm still brave!
And woodman leave the spot;
While I've a hand to save,
Thy axe shall harm it not.[155]

The poem ends on a militant note of firm resolve: the woodman shall not impoverish the poet's community of tree, family, history, and wild bird. One could certainly read "Woodman, Spare That Tree!" looking for the poet's sneer—that's how I first interpreted it. But the more time I spend mumbling over its lines, the more I become convinced that the passion of the poet comes from the very real countermodern attachment that he has for that almost-human tree. It's a friend and a relative: it roots him to a past of human work and mutual aid, one that recalls that other tree, the apple tree in Boston, the one rooted in the faith that "all things and all creatures are bound together, and live and flourish together."

This sequoian lesson was lost when the Kaweahans were forced from their land. Mills started popping up in the area in the 1850s, and soon dozens of lumbering operations were hacking deeply into groves of the Big Trees.[156] The Converse Basin, reputed to be the finest grove of sequoias in existence, was felled of every single sequoia—except for one, the Boole Tree, which the superintendent of the Sanger Lumber Company nobly left standing so that he could name it for himself.[157] Most of the trees, after being blown apart with black powder, went to the mills, which together churned out tens of thousands of board feet of lumber per day, to be turned into shingles, shakes, fence posts, garden stakes, and shipping crates—but mostly sawdust.[158]

The colonists weren't perfect: the archives are filled with records of their withering internal debates over minute points of ideological order, and there's a cloying attention to doctrine suffusing their discussion. One can also worry apart the colonists' claims to equality: though they did name a tree after Clara Barton, and women did hold positions of power, they were often in the "women's sphere": education, crafts, and music.[159] I'm nearly positive that there were no African American communards, and certain that there were no Chinese or Mexican American members. It's also true that the Kaweahans, except for a brief mention of the etymology of their name, had nothing to say about the American Indians.[160] Though I don't think it's too much of a stretch to imagine that the Kaweahans would have condemned Sherman's solution to the "Indian problem," it's also fair to point out that the Kaweahans did nothing explicitly to stop the physical and cultural slaughter. Perhaps a graveyard silence and public ownership of the Giant Forest is for the best.

Still, I can't bring myself to condemn the communards, nor to feel that their loss has been society's gain, for they remind me that reimagining the landscape can revolutionize society, culture, and economics; that the fight for social justice must always involve the physical, nonhuman world around us; that it's always worthwhile to dream; that enchantment has its own political value. The Kaweahans were eutopians, and we critics of capitalism need more like them.[161] After all, capitalism has its *Gospels of Wealth* and shrugging Atlases, its Alan Greenspans and Thomas Friedmans, all of whom reassure us that an unbridled free market will always deliver to each one of us all the riches, the personal freedoms, and the lives of fulfillment that we could possibly want, at absolutely no cost, as long as we, unaided, tug on our own boots every morning, and leave the wealthy alone.

12

What is sequoia?

When I hiked through Sequoia National Park in the summer of 2005, I had no idea of its past, knew nothing of the American Indians who had called it home, or the naming debates, or the communards. I knew nothing of sequoia, but I did watch the sun rise from atop Mt. Whitney, and I did marvel at the Big Trees as I took in that great expanse of quietude. For such experiences, rare in this world, I suppose I owe thanks to Stewart and Zumwalt, the Southern Pacific Railroad and even the US Army. When, together, they worked to reserve the land as a national park, they preserved a certain kind of space, one of silence. We often think of quiet as the sonic partner to peace, but in the modern world, violence—the violence of melting glaciers and rising seas and mutating cells and flowing capital—as often as not comes silently, and as oppressively, as midsummer humidity. The quiet of the park is the peace born of a rifle shot, its powder-borne silence extending into the present as well as retroactively drowning out the voices of alternative histories.

What is sequoia? For some, it was an American Indian or a tree, a route to riches, a spectacle at which to gawk, a symbol of a nation's chosenness. It was a revolutionary prophecy or a reaffirmation of the rights of conquest. It was very nearly an actual state, Sequoyah, one run by and for the Five Civilized Tribes in the first years of the twentieth century, and despite the knotty problem of the "real" Sequoyah, the idea of a Cherokee Cadmus continues to be a source of pride for many American Indians, who, despite all of Tecumseh Sherman's best attempts, survived to tell their own tales. And despite the best efforts of the loggers to turn the trees into quick cash, they, too, live on.

What is sequoia? Naming, like writing, is an act of possessing, always flirting with the violence of erasure, a telling of certain stories at the expense of others. But naming, like writing, like dreaming, is also an act of being possessed. When we make our marks, we are marked in turn. We don't know what sequoia is; it's a 250-foot-tall question mark whose supernaturally long life—if not cut down, blown up, burned; if left untouched by passing Gales, they are as immortal as any living thing can be—defies the notion of linear time, a thing rooted in the past but existing in the eternal present, much like history. We don't know what sequoia is; we just have the stories clothing the tree. And though no history, no landscape can ever be free of violence, some are more just than others. We all, each of us, make our own stories.

I'm tempted to end my sequoian tale tragically, though hinting at the slim possibility of romantic redemption, with a quotation from

Kaweah's Burnette Haskell who, despite bitterness, poverty, a broken heart, and serious addictions to alcohol and perhaps cocaine, could not ever quite see his commune as a mistake: "Is there no remedy, then, for the evils that oppress the poor? And is there no surety that the day is coming when justice and right shall reign on earth? I do not know; but I believe, and I hope, and I trust."[162]

I could end it here, but I won't. My sequoias ask for more. Anyway, I've spent too much time with the Kaweahans to settle for stingy hope.

Instead, it's Emily Dickinson, that "rare flower," as the twentieth-century critic Lewis Mumford called her, whose delicate petals topple walls and seed genuine dreams of a greener living world, whose words best nurture the promise of Kaweah:[163]

> It's all I have to bring today—
> This, and my heart beside—
> This, and my heart, and all the fields—
> And all the meadows wide—
> Be sure you count—should I forget
> Some one the sum could tell—
> This, and my heart, and all the Bees
> Which in the Clover dwell.[164]

ENDURING OBLIGATIONS

> Ah well, enough of the skulking rhetoric. Before we go we will plant a tree. I cleared away some ragweed yesterday, dug a thigh-deep hole this morning, and planted a young budding cottonwood this afternoon. We soaked the hole with well water, mixed in the peat moss and the carefully set-aside topsoil, and lowered the root ball of the sapling into its new home. The tree shivered as I packed earth around its base. A shiver of pleasure. A good omen. . . . We ourselves may never see this cottonwood reach maturity, probably will never take pleasure in its shade or birds or witness the pale gold of its autumn leaves. But somebody will. Something will.
>
> EDWARD ABBEY, *Down the River*[1]

The first elm tree I ever did see grew along University Avenue, in Fargo, North Dakota. The irony wasn't lost on me, an Easterner, that I had to head to the treeless Great Plains to walk down an elm-shaded street, and I dearly wish that it had been an idyllic experience; but the roar of gasoline-powered compressors, each connected to a single American elm via IV lines through which life-preserving chemicals were being pumped, made the experience rather more like strolling through an industrial chemotherapy ward than along an Arcadian boulevard. I was twenty-four. Those elms were dying thanks to the steady work of Dutch elm disease, a disease that spread along international trade routes, and so whose vector was wealth. The scourge first broke out in the United States in 1930, introduced by imported exotic lumber and the packing crates protecting luxury dishes; by 1933, as dust storms blasted the southern plains and gaunt bread lines stretched along city blocks, America's elms were dying in droves—seventy-seven million by 1980. It was an ecological calamity every bit as defacing as the Dust Bowl.[2] "An Elm can scarcely grow to old age without collecting rich human associations around it," wrote the sylvan biographer Donald Culross Peattie. But without their wooden witnesses, whatever these landscapes meant has blown away on a dry-coughing wind, and today's Elm Streets and various Elm Cities are now consecrated to ghosts.[3]

The dying elms are in good company: ash trees are withering across the Midwest, East, and South thanks to the emerald ash borer, a stowaway that likely also came in packing crates, while balsam fir, the Adirondack tonic for consumption, is in danger of a vast die-off, thanks to the practice of turning the northern woodlands into vast monocultures of the few species most desired by the construction market, species beloved of the budworm, which kills the balsam's skyward growth. Red spruce is right now melting away before the onslaught of acid rain. Eastern dogwood has been ravaged to the edge of extinction. And way out in the Golden State, Ponderosa pines are dying by the grove because of what was once mysteriously called California X-disease—but is really just California smog. Gypsy moths are denuding the forests of Michigan, and chestnuts, once a mainstay of East Coast forests and historically one of the most important trees to the American economy, have effectively vanished: three to four *billion* of them had died by the mid-twentieth century from an imported fungus.[4]

However bad this all is, we would be lucky if trees were the only things that have vanished from the modern American landscape. But it seems as if we are now in the middle of a global extinction that may be one of the greatest die-offs in the history of the world. This sixth extinction, as Elizabeth Kolbert makes clear, is entirely anthropogenic.[5] Indeed, intellectuals of all sorts, from geologists to critical theorists, are rushing to declare our geologic age the Anthropocene, or the age when human pollution, and particularly carbon dioxide, has assumed the sort of power long reserved for the earth and its nonhuman, natural processes. The Anthropocene, in other words, is the ultimate end of nature as a thing independent from humans, an era in which the United States has long played an outsized, defiant role as one of the world's biggest polluters.[6] And though we watch as island nations drown, as drought increases, as weather patterns become increasingly deranged, there is as yet no serious structural effort, among any mainstream political party in the United States, to take even the most basic of ameliorative steps. Instead, fracking for natural gas, mountaintop removal for coal, deepwater drilling for oil, and surface mining for tar sands has meant that production and consumption and environmental destruction and social erosion tick ever upward like the long-term graphs of the Dow Jones. And then, in 2016, a man who thinks climate change is a Chinese hoax found himself president of the United States.

Modernity, however, doesn't just destroy everything it touches; it also creates, and today in Nature's Nation you can *buy* an American elm: they are making a modest comeback thanks, in no small part, to the DuPont Corporation (purveyors of the cancer-causing defoliant Agent Orange) and its chemicals, as well as the genetic engineering of a patented, disease-resistant strain called American Liberty, a tree wholly owned by the Elm Research Institute.[7] There is, of course, an environmentalist modernity, and the one most Americans know well is the one of truly magnificent national and state parks, totaling nearly a 100 million acres of land, 108 million acres of wilderness, and 322 million acres of national and state forests and grasslands, a landscape watered by thirteen thousand miles of protected rivers.[8] We have mainstream environmentalism to thank for such everyday aspects of our life as curbside recycling, cloth shopping bags, cleanish air and clearish water, carbon offsets, organic food in the supermarket, hybrid cars, and lanes for commuter bikes. This is all a kinder side of modernity, and yet it's not at all certain that vacationing in Yosemite or shopping at Whole Foods — for those with the means to do so — can inoculate one against the mutagenic soil circling the homes of the one-in-four Americans who live next door to a Superfund site, or downwind and downstream from the excreted by-products of nuclear weapons tests and atomic energy plants and munitions factories, poisons all impossible to contain.[9]

Genetically engineered trees, preserved patches of woodland, and virtuous retail chains all may seem like accommodation in the face of global threat—and, of course, they are. As one of the biggest of the "big green" environmental organizations, the Nature Conservancy, puts it, "We pursue non-confrontational, pragmatic, market-based solutions to conservation challenges."[10] There's nothing at all odd about the Nature Conservancy's stance, either, for mainstream environmentalism has never offered much of a challenge to the modern order of things; instead, the very strength of environmentalism has long been its appeal to dominant social and economic conventions.[11] For instance, the Nature Conservancy, like Conservation International, the Conservation Fund, WWF, the World Resources Institute, and, for a time, the Sierra Club, all had or have strong ties to the oil industry—it's one arrow in environmentalism's quiver of market-based solutions to conservation challenges, a strategy that has helped ensure the tremendous success of environmentalism, including, most notably, its incorporation, beginning at the end of the nineteenth century, into every aspect of our politics, from local zoning boards to the Federal Environmental Protection Agency.[12] Yet such corporate influence came at the cost of a structurally unsettling, free-ranging green imagination. After all, it was in the closing decades of the nineteenth century that the work of a surveyor, writer, poet, political theorist, inventor, squatter, natural scientist, walker, and activist like Thoreau ceased to make realistic sense. One could be an environmentalist, or a social activist, but not both, and the recent rise of environmental justice helps underscore just how little justice has historically meant to environmentalism.[13] Instead of Thoreau's multifaceted radicalism, mainstream American environmentalism has followed the lead of Theodore Roosevelt—a man dedicated to wilderness and whiteness and wealth and martial manliness and the market—the nation's "eugenicist in chief," as Jared Farmer has put it; a man who counted John Muir as one friend, and Gifford Pinchot (the founder of American forestry) as another, and who was testing himself in the forever-wild

Adirondacks when he received word that President McKinley had been assassinated, and that he, Roosevelt, vice president of the United States, would become the new commander in chief; whose most influential book, *The Winning of the West* (1889), begins, "During the past three centuries the spread of the English-speaking peoples over the world's waste spaces has been not only the most striking feature in the world's history, but also the event of all other most far-reaching in its effects and importance"; whose presidency established national parks, championed scientific forestry and the US Forest Service, and helped make environmentalism popular by demonstrating that neither conservation nor preservation would ever challenge the dollar's Progress.[14]

And so the American witness tree could be cast as a last tragic landmark mutely testifying to a historical narrative that for all the world looks like a headlong rush toward a fate worse than death: one of the great ironies of Progress is that, rather than leaping forward, it seems to run time backward to when all life lived in a few puddles of primordial soup. "Modern society," writes Murray Bookchin, "is disassembling the biotic complexity achieved by aeons of organic evolution."[15] One of the great ironies of the history of modernity is that in seeking to set us free from the world of things, we've instead become their appendages— "producers" whose humanity has been demoted to our merest appetite, our need to consume. One of the great ironies of mainstream environmentalism is that despite its resounding success in protecting the land, it has offered only ineffectual rhetoric to the worldwide heating of Earth's atmosphere. It's a final irony that in divorcing ourselves from nature so that we might control it, we face a world beyond the reckoning of even our most advanced computing tools. Of course, there is a social dimension to all of this, and though each of us will shoulder a share of the cost of Progress, it is, as always under a capitalist ethos, the meek who shall inherit the trash heap.

And so this book, too, is perhaps nostalgic and irrelevant, a tragic history of losers and their dead landscapes, which, added together, signify exactly nothing. After all, Thoreau had no disciplined followers to disperse the seeds of his willowy thought, and it wasn't until the 1960s that he came to be seen as something more than one of Emerson's minor students.[16] The Adirondack pioneers, like the Kaweah communards, have been so thoroughly forgotten that their stories, if told at all, are mere footnotes to triumphal tales of John Brown and park building, while the supple and complicated history of wilderness, once the cure for consumption, has calcified into ahistorical irony. Out West in Utah, the skeleton of A. J. Russell's Thousand-Mile Tree was finally removed in 1900; it was a new century, after all, and the tree stood in the way of Progress: it was cut down to run the railroad's track straighter towards the Golden State, where, in 1937, just past the left-hand turn to eutopia, the giant sequoia named Witness Tree was also felled.[17]

But enough of the skulking narrative. Decline and Progress, like

FIGURE 43. © Frank Gohlke, *Landscape, St. Paul, Minnesota.* 1974. Courtesy, George Eastman Museum.

destruction and creation, are two sides of the same narrative coin, after all.[18] The world may be damaged; it may, in fact, be broken. But it is ours, the only one we've got.[19] "The sun shines to-day also," wrote Emerson, and the great solar strength of countermodern culture has always been its generative diffusion, its refusal to remain focused around just one political, intellectual, or cultural tradition, its ability to fuse seeming incommensurables into new things startling in their unexpectedness, its faith in the immortality of ideas.[20]

In 1956, Allen Ginsberg's human *Howl* rent a decade's quiet desperation: "I saw the best minds of my generation destroyed by madness, starving / hysterical naked," run the famous first lines; but it is another, less obvious of his poems that I think best captures an always-present spirit of American cultural countermodernity.[21] Entitled "America," it begins:

> America I've given you all I have and now I'm nothing.
> America two dollars and twentyseven cents January 17, 1956.
> I can't stand my own mind,
> America when will we end the human war?
> Go fuck yourself with your atom bomb.
> I don't feel good don't bother me.
> I won't write my poem till I'm in my right mind.
> America when will you be angelic?

"America" is, above all else, a patriotic poem, and he dares the better angels of our human nature to come out from hiding, to make peace. It's also a dream of a landscape yet to be realized, of a world worth inhabiting, one that honors the legacies of the Wobblies, the Spanish Loyalists, Sacco and Vanzetti, the Scottsboro boys, the American communists and American Indians and African Americans and queer Americans. Enough of the war against humans, he writes, enough of the cold war against humanity: "It's true I don't want to join the Army or turn lathes in precision parts / factories." And though he lobs the writer's ultimate suicidal threat—to silence his own pen—in the end, no matter how rotten his stomach feels, no matter how much beauty and creativity and knowledge an insane America has turned into mutually assured destruction, the world, the wild world, the world of Ginsberg's "mystical visions and cosmic vibrations," enchants.[22]

At nearly the same moment, Rachel Carson was working on her masterpiece, *Silent Spring* (1962), whose second chapter she entitled "The Obligation to Endure." If Ginsberg howled the countermodern spirit, it was Carson who put her nib on the beating heart of countermodern *green* thinking: a tenacious love of defiantly enchanted life, rooted always in both the land and the human mind. "The 'control of nature,'" Carson wrote in her still-resonant final paragraph, "is a phrase conceived in arrogance, born of the Neanderthal age of biology and philosophy, when it was supposed that nature exists for the convenience of man."[23] Carson never explicitly addressed the Neanderthal age that supposes some humans exist for the convenience of others (though in her chapters on industrially caused birth defects, she implied as much). But Murray Bookchin did. His first book, *Our Synthetic Environment* (1962) debuted five months before *Silent Spring*, and its key insight was that environmental issues and social issues are ultimately expressions of the same things—violence, exploitation, hierarchy—and so cannot be disentangled from each other. This is an insight that has just recently found itself reincarnated in the pages of Naomi Klein's *This Changes Everything: Capitalism vs. The Climate* (2014), which locates the loosing of climate change in the Pandora's box of what Klein calls "extractivism" but is really just another name for a culture of all-out exploitation. One of the many strengths of Klein's book is that it refuses to remain a critique, and instead looks forward to the coming day when the split is repaired between nature and culture, between human and human, between environment and justice. But for that reconciliation to happen, "a worldview," Klein writes, "will need to rise to the fore that sees nature, other nations, and our own neighbors not as adversaries, but rather as partners in a grand project of mutual reinvention."[24]

Environmentalism has its place, and I celebrate its successes. But an environmentalism without music or poetry or an explicit commitment to justice doesn't offer our world much beyond land preservation, industrial efficiency, and beer-can recycling. That's not enough. If mainstream environmentalism seems unfit for the challenges symbolized by global climate change, it's because such challenges have never been simply problems of the environment, but of modernity, problems every bit as intellectual and cultural and social as environmental. We need Ginsberg's poetry, and Carson's advocacy, and Klein's mutual-aid inspired journalism; we need the intellectual capaciousness of landscape; and we need history, for though Klein is right that the old extractivist mindset is long obsolete, we don't have to reinvent its replacement out of thin air. We have the past for inspiration, can breathe in its spirit to awaken, enchanted, from the present's sterile passivity.[25]

Thoreau and A. J. Russell may be long dead, but their work survives; and though the communities planted amongst the sequoias and the Adirondack pines are rarely remembered, they all gesture together to a much wider cultural legacy, for countermodern wonder is nearly anywhere one looks. For instance, even as the Kaweahans were busy renaming their forest trees, a trio of intellectuals, Henry George, Henry Demarest Lloyd, and Edward Bellamy were honing philosophies with keen social and environmental edges, ones especially suited to the growing cities and factories of the United States.[26] Their influence spilled into the twentieth century and resonated with a continuing legacy of vibrant critical and radical landscape visions: in the cities of Lewis Mumford—who wrote in defense of good places, "If our eutopias spring out of the realities of our environment, it will be easy enough to place foundations under them." It also spilled into the criticism of Waldo Frank and the beloved community of Van Wyck Brooks, into the urban anarchism of Emma Goldman, into Carey McWilliams's critiques of industrial agriculture, into the human rights advocacy of Cesar Chavez, into the way Ralph Ellison's *Invisible Man* shined 1,369 light bulbs on the hidden landscapes of racism.[27]

The activism continues in our contemporary United States, especially around the environmental fringes in the off-beat collectives and small gatherings devoted to remaking society and environment at the same time: the communards, the do-it-yourself mechanics converting the cast-off Mercedes of the nation's rich to run on used vegetable oil, the appropriate technology folks, the People's Climate Marchers, the vegetarians and vegans and fruitarians and locavores, those who support community and urban agriculture, the bicycle evangelists, the Occupiers, Water Protectors, and tree sitters—like Julia Butterfly Hill, who lived in a 1,500-foot-tall redwood for 738 days because she knows that a redwood has a higher purpose than someone's pool deck.[28]

It's there, too, in music, in Woody Guthrie's Thoreauvian determination, once he left the redwood forest, to roam and ramble and follow his footsteps wherever they might take him—

> There was a big high wall there that tried to stop me
> A sign was painted said: Private Property,
> But on the back side it didn't say nothing—
> This land was made for you and me[29]

—or more recently in indie-pop sensation The Arcade Fire's 2010 album *The Suburbs*, which culminates (for me) in the haunting lines of their song, "Half Light II (No Celebration)": "Oh, this city's changed so much since I was a little child / Pray to God I won't live to see the death of everything that's wild."[30] It can be seen on film in Pare Lorentz's *The Plow That Broke the Plains* (1936), Spike Lee's *Do the Right Thing* (1989), and Jim Jarmusch's *Dead Man* (1995), and stilled in the photographs of Peter Blake's furious photo essay, *God's Own Junkyard: The Planned Destruction of America's Landscape* (1964) or Trevor Paglen's sublimely terrifying *Invisible: Covert Operations and Classified Landscapes* (2010).

Of course, a vibrant green pulse thrums its way through American letters, and not only in the expected genres: nature and environmental writing. The green light in F. Scott Fitzgerald's *The Great Gatsby* (1925) has enchanted us for nearly a century, while the industrial-pollution-scarred population (including the wheelchair-bound hit men who lost their legs jumping in front of trains) living in the Great Concavity of David Foster Wallace's *Infinite Jest* (1996) has been perplexing us for twenty years. It's there in the tales of enchanted forests and magical animals that we read to our children in the hope that landscape will teach them how to lead good lives, and it's there in the essays that we read to our spellbound selves deep into the early hours of the morning, from

E. B. White's *This Is New York* (1949), which finds under the shade of a city tree the dogged refusal to succumb to "the cold menace of human suffering," to Rebecca Solnit's *A Field Guide to Getting Lost* (2005) and its remarkable faith that a home of sorts always lies in the blue of distance.[31]

"What more can you do? What more can you do?" asks the poet, environmentalist, and antiwar activist W. S. Merwin in his *Unchopping a Tree*.[32] I won't continue anymore to hack out the outlines of a persistent, resistant American green culture; it grows everyday more rapidly than my words can define, it's all around, and besides, this book was never about binding the past between the leathern covers of tradition. The point is simply this: business culture didn't triumph in the nineteenth century, nor has it today. What has triumphed is the story that all American culture is business culture, that free markets and free humans go hand in hand, that conquering space and time has enthroned us all, that the world is made of inert matter free for the taking, that we can invent and buy our way to a better tomorrow. Those are the stories of the already dead. We who live today need something better.[33]

* * *

Writers sometimes like to say that all writing is autobiographical, and if that's true, then this history is an intellectual autobiography of my attempt of nearly fifteen years to look one short paragraph of writing squarely in the eye. I can remember, vividly, when I first encountered it, back in the beautiful fall of 2003. I was teaching history and environmental issues at an alternative high school on the Maine coast, feeling a bit like an impostor: I was not, am not, will probably never be, an environmentalist. I've always been more interested in cultural and intellectual radicalism than a movement that first manifested itself to me as suburban, privileged, white, and therapeutic; yet there I was, teaching environmental issues at a school that included an organic farm and whose mission was to promote environmental awareness. Being bookish, I figured that I ought to do a little reading, and so on October 18, I walked into the Gulf of Maine bookstore in Brunswick, Maine, and, for $7.34, bought myself a copy of Edward Abbey's *Desert Solitaire*.

Right in the very beginning of the book, at the end of the author's introduction, comes the part that won't let me go. "Most of what I write about in this book is already gone or going under fast," Abbey writes: "This is not a travel guide but an elegy. A memorial. You're holding a tombstone in your hands. A bloody rock. Don't drop it on your foot— throw it at something big and glassy. What do you have to lose?"[34] Those six sentences gave birth to the thing you're now holding in your hands, and I still feel the bottom drop from my stomach when I read the audacious call to action, the riffing on Marx and Engels's "nothing to lose but their chains," the despair hovering on an angry hell-with-it-all nihilism—those sentences are a dare, "what do you have to lose" both a question and a statement, one that I've always been afraid to accept. I have radical friends—anarchist homesteaders, dumpster divers, EMTs who volunteer to treat the protesting victims of police brutality; my wife, Talia, played an active part in the 1999 Seattle World Trade Organization protests and West Coast tree sits. But I wrote every word of this book first in the basement of a glorious building on the campus of a genteel Ivy League university, and later in a three-windowed, air-conditioned office at one of the Midwest's premiere public schools hoping to earn my way into the comfortable life of a tenured college professor.

Nevertheless, Abbey's words worked their way into my DNA, and in October 2010, I was compelled to reread *Desert Solitaire* for the first time, a book that I now approached tentatively: I was afraid that maybe it wouldn't appeal to me the way that it had when I was twenty-three. And I was afraid to confront the level gaze of the author's introduction, asking me what I had done with the past seven years.

What *had* I done? A few months earlier, in August, Talia was ten weeks pregnant, and we went to her midwife's office, where the midwife pointed some sort of mechanical ear at Talia's barely rounded belly. We then heard the heart beat—rapid and perfect—of the person who would turn out to be our first child, our son, Wyeth. Though I have had the rare good fortune in my life to be immersed in beautiful music and rich conversation, it all faded to static behind the sound of our healthy baby's beating heart.

One day later I was diagnosed with testicular cancer. Two days after that I was whisked to the operating table, and the abominable life that had taken root in *my* body was cut out. But not entirely, and that is how I found myself, in October, back with *Desert Solitaire*, exactly seven years after I had finished it, in the "infusion suite" at the local hospital, watching nurses too expertly insert IV lines into my chest in preparation for six weeks of intensive chemotherapy, a period during which I lost my hair, and for a while some of the feeling in my fingertips, and months when I should have been supporting Talia, reading up on fathering books, preparing the house, dreaming of a future of diapers and laughing and first steps. I lost my fear of needles. I lost faith in my body. And I now sport surgical scars and weird blotches on my skin left by one of the drugs in my three-drug chemo "cocktail."

I was thirty, and Abbey, from between his covers, asked me, "What do you have to lose?"

As I revise this, it has been five years since the last round of chemo mercifully ended, and the cancer is gone. I'm considered cured. But I still think about cancer every day, every time my mind wanders to the recent past, or the unknowable future, every time I go for a run, every time I get undressed and find myself confronted by marred skin. On the bad days, I lie awake at night, once my family has gone to bed, acutely aware of the aches that happen to fall anywhere between groin and shoulders—as they almost inevitably must—petrified as the black fear comes gnawing its way back. I think of those dying elms back in Fargo.

"What do you have to lose?" asks Abbey from the dark.

There were an estimated 1,529,559 other new cancer cases in 2010. Between 2004 and 2006, the probability that any American man, of any age, selected at random, had cancer was nearly one in two. Women had a one in three chance.[35] And so I am a naturalized member of the ever-expanding Kingdom of the Ill, as Susan Sontag—who beat cancer twice, until it overtook her a third and final time—put it in *Illness as Metaphor*. "Everyone who is born holds dual citizenship," she wrote, "although we all prefer to use only the good passport, sooner or later each of us is obliged, at least for a spell, to identify ourselves as citizens of that other place."[36] Thankfully, the death rates for cancer in the United States have been dropping—twenty-five years ago mine probably would have been fatal; yet even as fewer people die, an increasing number wake up one morning to find a lump where once was smooth skin, for cancer rates are on the rise worldwide, particularly in industrialized, modern countries. Sontag warned against using cancer as a metaphor, both because linguistic tricks tend to make cancer shameful for those of us who are given it, and because she was afraid that using cancer as a metaphor would incite violence: we, after all, still speak of the war on cancer, and bombarding our tumors with radiation, and weakened bodily defenses, as if we've all been unwillingly drafted into the military's killing machine. But I think Sontag erred too far on the side of caution, because cancer is more than a metaphor. It is the flesh-and-blood body born in the image of modern capitalism, a self-interested actor seeking to maximize its own benefit, and so win at the great race of life by growing endlessly while all the other billions of cells in our body plod along socially, contentedly, healthfully.[37]

Rachel Carson, who died of breast cancer, noted that the dystopian possibilities of nuclear holocaust were only slightly more worrying—and, in retrospect, it turns out, less probable—than the prospect of our environment's total contamination at the hands of industry.[38] That contamination has increased markedly since Carson's death. Which means that my son, my beautiful, perfect little boy who has never done anything to anyone, who wants nothing more than to play with his trains and eat popsicles or pasta and dance joyously to our rockabilly records—he will have a far greater chance of contracting cancer than his father.

"What do you have to lose?" asks Abbey.

My cancer has nothing, in the end, to do with me: no history of testicular cancer, no preconditions, nothing, save for the fact that I, like all of you, am being knowingly poisoned for someone's profit. The World Health Organization has estimated that at least 80 percent of all cancers stem from "environmental" causes—the preferred term that really means "industrially-produced toxins."[39] Some carcinogens, like dioxin, like DDT, like benzene, are so prolific that they are thought to inhabit the tissues of every single living person in the United States. In the mid-1990s, the writer and scientist Sandra Steingraber—a bladder cancer survivor—wrote that cancer researchers estimated somewhere between 3,750 to 7,500 chemicals in daily use were carcinogenic, though because there's far more political and economic will to produce moneymaking poison than sequester it, only two hundred of those suspected carcinogens had been tested and regulated.[40] It turns out that capitalism, though it may not be very good at ensuring wealth or equality or justice for anyone at all, is an incredibly democratic scheme for spreading pestilence everywhere and to everybody. This is modern Progress at the beginning of the twenty-first century, in the wealthiest nation in the world.

As I fearfully look ahead, waiting for the day to come, as statistically it must, when my oncologist once again enters the exam room with his practiced and polished stone-faced mask, I am sorely tempted to finally accept Abbey's challenge and urge you to throw this book, as hard as you possibly can, against all those big and glassy things where lives are destroyed for profit: the corporate boardrooms and university labs where chemicals and petroleum products are dreamed up, bought, and sold; the military bases where they are perfected into death-dealing devices; the political chambers where their toxicity is routinely downplayed in favor of profit—the whole Kingdom of the Ill and its ruling aristocrats: cancer, global climate change, industrial pollution. I want to ask you to throw this book at anything or anyone that makes a dollar off of your misery, that separates living things into the valued and the valueless; I want to dare you to use these pages to smash any hierarchy that keeps you from living the healthy, fulfilling life that is yours by birthright.

But I can't. Though my own body has certainly been damaged, and it may, in fact, be genetically broken, it is mine, the only one I have. Ideas may be weapons, but I'm tired of violence, and I refuse to live, or raise my son, or share my book with you in a deconstructed world of broken glass and shattered concrete. Ideas may be weapons, but they make better seeds.

Though Abbey asks us to imagine *Desert Solitaire* as a tombstone, I've had enough of death. And it's exactly here on despair's thin frontier that I finally understand this ghost who has been whispering in my ear. No one writes a book, a book about beauty, a book whose sylvan epigraph ends "Love flowers best in openness and freedom"; no nihilist writes a book like *Desert Solitaire*.[41] I think I finally discovered what Abbey had to say as I sat in the chemo ward trying not to fixate on the distended bags of chemicals, with their warnings of biohazardous material, seeping into my damaged body while the IV pump whirred like a mechanical cicada.

Love flowers best in openness and freedom. In 2015 we welcomed another son, Everett, into our family, and like his older brother, he is healthy, and beautiful, and perfect.

Photographs and poems and essays and maps all grow from fertile ideas; they're like trees that someone once planted, and what is a planted tree if not a gift from the past to the future? This book, then, is no memorial to the dead, no vengeful battering ram, but a green, enchanted, ungovernable, wild-talking tree, a witness tree rooted in a living history, a thing under whose madly chattering leaves we can all gather, like Thoreau advises, "as *eu*peptics to congratulate each other on the ever glorious morning," a place of laughter and friendship and music and poetry—our own balsam-scented eutopia—just one of the many places where, as we wait for the next tale of decline, we can do what we must to build a world of our own.[42]

What do we have to lose?

ACKNOWLEDGMENTS

I grew up in the woods and farm fields of rural upstate New York, a child of the country with a predilection for wandering, and though it is hard for me to pinpoint the exact birthplace of this book, it is easy enough to find the muddy, curious boot prints of my upbringing on the following pages. We didn't have much money, but my parents, Maggie and Eric, always made sure that our house was full of books and music and art and lively discussion, and always, always lots of love, especially when times were tough. I have two kids of my own now, and if I'm doing any good at all as a parent, it's due to the example set me by my own. Carrie, Sam, and Hannah Johnson have also always been there with warm hearts, open arms, and soul-comforting food.

Had I not taken a job as a history and environmental issues and ethics teacher at the Chewonki Foundation's Maine Coast Semester, there would never have been a book to write. At least not this one. Scott Andrews and Sue West, Bill Hinkley, his wife Amy, and their towheaded boys Max, Ezra, and Amos gave me a glimpse of what commitment could make possible. Jesse Pine Dukes has become a travel companion, drinking partner, and intellectual opponent whose friendship I deeply treasure. And then there's Paul Arthur, my co-teacher, confidant, and agent provocateur who, more than any other single person I know, has shown me why a life of the mind is worth working, sacrificing, and at times fighting so hard for.

When I left Maine for Cornell I found four mentors who would spend the better part of the next decade supportively and continuously training, pruning, trimming, and ultimately encouraging my intellectual growth to take on its own unique form. It's hard to remember a time when I didn't look out at the landscape through eyes made more perceptive by Andrea Hammer, or more political by Ray Craib, who first introduced me to Henri Lefebvre's *The Production of Space*. Ron Kline was a thrilling guide to the terra incognita of nineteenth-century science. And Aaron Sachs—well, it's hard to overstate the influence Aaron has had on nearly every aspect of my intellectual development, from writing to parenting. Aaron exemplifies the care, the commitment, and the stubborn independence that I have long associated with the best in academia.

Cornell University's history department—a place that typifies intellectual curiosity—was the perfect place for me to pursue a PhD. I got to serve as one of Nick Salvatore's teaching assistants and watch daily as one of the great historical minds of our age conjured the past into the present. The cohort gathered around the Americanist colloquium started by Ed Baptist and John Parmenter was the single most intellectually stimulating departmentally organized aspect of my time in graduate school. I was also fortunate to be a member of a half dozen writing groups all filled with brilliant, passionate scholars: Daniel Ahlquist, Tom Balcerski, Angie Boyce, Dianne Cappiello, Duane Corpus, Mari Crabtree, Brian Cuddy, Elisa DaVià, Ryan Edwards, Brigitte Fielder, Will Harris, Toni Jaudon, Maeve Kane, Candace Katungi, Peter Lavelle, Max McComb, Vernon Mitchell, Trais Pearson, David Rojas, Melissa Rosario, Djahane Salehabadi, John Senchyne, Mike Schmidli, and Rebecca Tally, among many others.

There is one group that deserves special mention: Cornell's Historians Are Writers! Heather Furnas, Rebecca Macmillan, Katie Proctor, Aaron Sachs, and I all settled down to the first meeting of HAW! back in 2007. The group was founded on the principle that literature and academic writing should not be antithetical propositions: to the contrary, the best history has always also been, and must continue to be, literary (hence the exclamation point). Thanks especially to Sarah Ensor, Melissa Gniadek, Tina Post, Rob Vanderlan, and Josi Ward, all of whom read drafts of my work.

There are a few folks who deserve special mention on their own, the sorts of people whom I left Maine hoping to find. Foremost among these is Amy Kohout, fellow HAW!er, about whom I could write paragraphs. Suffice it to say that she's the sort of person who knits you a lap blanket for your wedding, a chemo cap for when you fall ill and, when you're finally healthy again, hammers you with intellectual provocations over home-cooked tomato soup. For all the snarky caricatures of the ivory tower, it's the only place where one could meet someone like Franz Hofer. Introduced to each other by a shared interest in punk rock, we soon bonded by sampling some of the world's finest beers and later forged intellectual bonds discussing photography and visual culture—I'll long remember our discussion of *Let Us Now Praise Famous Men*. Brent Morris and I cheered for cars as they slid down impossibly snowy Ithaca streets, mountain biked in Shindagin Hollow, and rode around Cayuga Lake on the Fourth of July. Brent's also a phenomenal historian whose deep knowledge of abolitionism astounds me. Katie Proctor wrote me the single greatest, fist-in-the-air-pumping, life-affirming e-mail that I have ever received: "We've got work to do." Charis Boke, and her determination to care, remains an inspiration, and I miss our meetings in dimly lit, Dostoyevskian teahouses, there to read radical literature and plot revolution. Ben Wang gave me a hand when I really needed one. I'm not sure how many miles we ran together, but Ithaca will always remind me of long, early morning runs on freezing-cold Saturdays in February and March. And the truly brilliant poet, historian, and restoration ecologist Laura Martin read and commented upon every line of an early draft of this book—an act of heroism. She's been a good friend in dark times and the best person I know with whom to spend an evening talking about writing.

A tremendous thanks also to my doctors: Dr. Sami Husseini, Dr. Charles Garbo, Dr. Douglas McNeel, and especially the nurses at the Cayuga Cancer Center whose kindness got me through the worst.

I've been lucky. My research has been supported by a number of fellowships and awards: two Sage Fellowships from Cornell University; two research grants from Cornell's American Studies Program; a summertime Walter LaFeber Fellowship from Cornell—which allowed me to find A. J. Russell; a Dissertation Proposal Development Fellowship, funded by the Andrew W. Mellon Foundation and administered by the Social Science Research Council (one of the most intellectually stimulating gatherings I've ever had the good luck to be a part of; thanks especially to Sumathi Ramaswamy and Martin Jay, the brilliant Christine DeLucia, and the deeply caring Tom Okie, who also read an early version of my manuscript); a Boldt Fellowship from Cornell; a Jay and Deborah Last Fellowship at the American Antiquarian Society/Center for Historic American Visual Culture; a Sustainability Grant from Cornell's Society for Humanities Initiative for Sustainability via the Humanities and Arts; and the Hampel Award from Cornell. A very generous book subvention award from the Ralph Waldo Emerson society went a long way toward making this book as visually stunning as it is.

For two years I wrote at the University of Wisconsin–Madison as an A. W. Mellon Postdoctoral Fellow in the Humanities. It was an invaluable gift that allowed ample time for reading, thinking, and revising. I'd like to thank Jess Courtier, Sarah Guyer, Susan Friedman, Sarah Florini, Amanda Rogers, Brian Goldstein, David Zimmerman, Mari Lepowski, Rob Howard, Jill Casid, Will Jones, Bill Cronon, Jolyon Thomas, Elaine Fisher, Darryl Wilkinson, Emily Clark, Marrion Ladd, Megan Massino, Yi-Fu Tuan, Sarah Dimick, Ann Harris and Max Harris, Liz Hennessey, Bethany Moreton, and Pam Voelker. Gregg Mitman welcomed me to Madison, invited me join the Center for Culture, History, and Environment, and was relentless in encouraging me to write, write, write. Jerome Tharaud is a model of commitment—to literature, to fast mile paces, to the deep humanity of thinking, to family—and I'm lucky to call him a friend. Finally, my Madison postacademic writing group of intellectual risk takers—Nathan Jandl, Mary Murell, Andrew Kay, Heather Swan, and Adam Mandelman—has been a godsend.

Nice as all of the funding has been, without support of archivists and research staff throughout the country, this book would never have gotten off the ground. I'd like to thank the folks at the Concord Free Public Library; the Miriam and Ira D. Wallach Division of Art, Prints, and Photography at the New York Public Library; the Huntington Library; the American Antiquarian Society and the Center for Historic American Visual Culture; the Morgan Library and Museum; Yale's Beinecke Rare Book and Manuscript Library; the Adirondack Museum; the Bell Memorial Library; Cornell's Division of Rare and Manuscript Collections; the University of Wisconsin–Madison Library; the Wisconsin Historical Society; and the Special Collections Research Center at Syracuse University. I'd especially like to thank Leslie Perrin Wilson, the curator of special collections, at the Concord Free Public Library. I hadn't intended to spend any time looking at Thoreau's Concord River Survey, and it was only at her insistence that I unrolled its seven-foot length and stood astounded. Also, the American Antiquarian Society is a paradise for pre-twentieth-century Americanists, and I would especially like to thank Paul Erickson, Gigi Barnhill, Lauren Hewes, Laura Wasowicz, and the AAS's pages for ensuring that the fruits of my research were especially bountiful.

Phil Terrie, dean of Adirondack history, helped immeasurably with Act Two, and Russell scholar Glenn Willumson—who I've never met in person—sent me a number of documents on A. J. Russell that I had been unable to find. Terrie and Willumson exemplify everything that is good about academia.

Levi Stahl, Nora Devlin, Rachel Kelly, Isaac Tobin, Katherine Faydash, and Yvonne Zipter at the University of Chicago Press have been extraordinary in their efforts to make this book everything it is. I also owe a debt of gratitude to Jared Farmer, Richard Higgins, and the anonymous reviewers who helped me sharpen my thinking and my prose. But it's really my editor, the fellow traveler Tim Mennel, who did more than anyone else to draw out my voice and style. There's no one else I'd rather have reading my drafts, and though I thought the effort of revision was going to kill me, I wish a Tim for every writer out there. Be reasonable, demand the impossible, indeed.

Writing this book has brought me a long way from home, and not a single word of it would have been worth a thing were it not for my two perfect, soulful, feral little boys, Wyeth Gabriel, who came along just as I was finishing the first draft, and Everett Ellis, who has seen me revise and revise again. My beautiful, brilliant, strong Talia, the love of my life with the big smiling eyes, has been my companion as we roamed from the spruces and salt waters of Midcoast Maine to the prairies and burr oaks of Wisconsin: all the passion behind these words ultimately is for you. Sunshine daydream, forever.

ABBREVIATIONS USED IN NOTES

"History" "The History of the Kaweah Cooperative Colony of California. A Record of a Remarkably Successful Experiment of Fifty Years Ago. Revealing Also its Illegal and Ruthless Wrecking at the Hands of the Harrison Administration of 1889–93. A Statement of Facts Hard to Believe but Nevertheless Confirmed by Incontestable Documentary Evidence," 1937, MS, The History of Kaweah and Accompanying Papers, Yale Collection of Western Americana, Beinecke Rare Book and Manuscript Library
AHR *American Historical Review*
AM *Atlantic Monthly: A Magazine of Literature, Art, and Politics*
APB *Anthony's Photographic Bulletin*
AQ *American Quarterly*
BDE *Brooklyn Daily Eagle and Kings County Democrat*
CEP *Thoreau: Collected Essays and Poems* (New York: Library of America, 2001)
CHDT Walter Harding and Carl Bode, eds., *The Correspondence of Henry David Thoreau* (New York: New York University Press, 1958)
CS *Concord Saunterer: A Journal of Thoreau Studies*
CUPR *Photographs Taken during the Construction of the Union Pacific Railroad*, 1864–1869, Yale Collection of Western Americana, Beinecke Rare Book and Manuscript Library

ABBREVIATIONS USED IN NOTES

CW	*Commonwealth: A Journal for Those Who Labor and Who Think*
EH	*Environmental History*
ER	*Environmental Review*
ESQ	*ESQ: A Journal of the American Renaissance*
FN	Field Notes of Surveys made by Henry D. Thoreau since November, 1849, Henry David Thoreau Papers 1836–1862, vault A35, Thoreau, unit 1, box 1, folder 11, William Monroe Special Collections, Concord Free Public Library
FS	Bradley P. Dean, ed., *Faith in a Seed: The Dispersion of Seeds and Other Late Natural History Writings* (Washington, D.C.: Island Press, 1993)
GSP	Gerrit Smith Papers, Special Collections Research Center, Syracuse University
GWI	*The Great West Illustrated in a Series of Photographic Views Across the Continent; Taken Along the Line of the Union Pacific Railroad, West from Omaha, Nebraska. With an Annotated Table of Contents Giving a Brief Description of Each View; Its Peculiarities, Characteristics, and Connection with the Different Points of the Road*, vol. 1 (New York: Office, 20 Nassau Street, 1869)
GWW	*Geo. W. Williams & Co.'s Carolina Fertilizer with Twenty Photographic Views from the Line of the Union Pacific Railroad* (Charleston, SC: n.p., 1869)
HDTP	Henry David Thoreau Papers 1836–1862, William Monroe Special Collections, Concord Free Public Library
HDTS	Henry David Thoreau Surveys, William Monroe Special Collections, Concord Free Public Library
HFP	Haskell Family Papers, 1878–1951, Bancroft Library, University of California–Berkeley
J19	*J19: The Journal of Nineteenth-Century Americanists*
JAH	*Journal of American History*
JAM	Henry David Thoreau, Journal: Autograph Manuscript, April 8, 1859–September 21, 1859, MS. MA 1302.35, Morgan Library
JHDT	Bradford Torrey and Francis H. Allen, eds., *The Journals of Henry David Thoreau*, Dover edition (New York: Dover Publications, 1962)
KCCP	The Kaweah Cooperative Colony Papers, Yale Collection of Western Americana, Beinecke Rare Book and Manuscript Library

LT	*The Liberty Tree*, Chicago
MP	*Massachusetts Ploughman and New England Journal of Agriculture*
NE	*National Era*
NN	*Nunda News*, Nunda, New York
NS	*The North Star*
PA	*Photograph Albums of Utah, Wyoming, Nebraska, and California*, Yale Collection of Western Americana, Beinecke Rare Book and Manuscript Library
PP	*Philadelphia Photographer*
RWE	Mary Oliver, ed., *The Essential Writings of Ralph Waldo Emerson* (New York: Modern Library, 2000)
SPRM	*Sun Pictures of Rocky Mountain Scenery, with a Description of the Geographical and Geological Features, and Some Account of the Resources of the Great West; Containing Thirty Photographic Views along the Line of the Pacific Railroad, from Omaha to Sacramento* (New York: Julius Bien, 1870)
UPRRV	*Union Pacific R.R. Views* (Boston: A. Mudge, [1869?])
USM	*United States Magazine, and Democratic Review*
W	Carl F. Hovde, William L. Howarth, and Elizabeth Hall Witherell, eds., *A Week on the Concord and Merrimack Rivers* (Princeton, NJ: Princeton University Press, 2004)
WWMC	*Thoreau: A Week on the Concord and Merrimack Rivers; Walden; The Maine Woods; Cape Cod* (New York: Library of America, 1985)

NOTES

WHEN THE BOUGH BREAKS

1   John Prine, "Paradise." Words and Music by JOHN PRINE © 1970 (Renewed) WALDEN MUSIC, INC. All Rights Reserved. Used by Permission of ALFRED MUSIC. 296, 1971.
2   *The Old Elm and Other Elms* (Boston: A. Williams & Co., 1877) 2; "Read and Run," *MP* 35, no. 21 (Feb. 19, 1876), 2.
3   *Old Elm and Other Elms*, 5; *Memorial of Jesse Lee and the Old Elm: Eighty-Fifth Anniversary of Jesse Lee's Sermon Under the Old Elm, Boston Common, Held Sunday Evening, July 11, 1875, with a Historical Sketch of the Great Tree* (Boston: James P. Magee, 1875), 20, 51–52; "To the Old Elm in the Centre of Boston Common," *Boston Monthly Magazine*, June 1826, 53. For more on the history of the Great Elm, see Thomas J. Campanella's *Republic of Shade: New England and the American Elm* (New Haven, CT: Yale University Press, 2003), 48–54.
4   Or at least many in the nineteenth century assumed that the tree on Bonner's map was the Great Elm, though it's not clear that it actually was. My thanks to Anne Beamish and her paper "The Venerable Relic: The Great Elm on the Boston Common" (American Society for Environmental History annual conference, San Francisco, Mar. 2014).
5   "Westward the course of empire takes its way" comes from George Berkeley's poem "Verses on the Prospect of Planting Arts and Learning in America," and "the annihilation of space and time" comes from poet Alexander Pope's *Peri Bathous*; the latter reverberated throughout the nineteenth

century to be eventually picked up by Karl Marx. See George Berkeley, "Verses on the Prospect of Planting Arts and Learning in America," in *American Poetry: The Seventeenth and Eighteenth Centuries* (New York: Library of America, 2007), 346; Alexander Pope, *Peri Bathous; or, Martinus Scriblerus; His Treatise of the Art of Sinking in Poetry*, in *The Prose Works of Alexander Pope*, ed. Rosemary Cowler (Hamden, CT: Archon Books, 1986), 2:211.

6   Alexis de Tocqueville, *Democracy in America*, trans. and ed. Harvey C. Mansfield and Delba Winthrop (Chicago: University of Chicago Press, 2000), 461.

7   For more on how limitless growth came to be an unquestioned tenet of American culture, see Eric Hobsbawm, *The Age of Capital: 1848–1875* (1975; repr., London: Abacus, 1997); Christopher Lasch, *The True and Only Heaven: Progress and Its Critics* (New York: W. W. Norton & Co., 1991); and Steven Stoll, *The Great Delusion: A Mad Inventor, Death in the Tropics, and the Utopian Origins of Economic Growth* (New York: Hill and Wang, 2008).

8   The relevant historiography is far too lengthy for me to comprehensively cite here. For works particularly important for my own understanding of the nineteenth-century United States, see T. J. Jackson Lears, *No Place of Grace: Antimodernism and the Transformation of American Culture, 1880–1920* (Chicago: University of Chicago Press, 1981); Alan Trachtenberg, *The Incorporation of America: Culture & Society in the Gilded Age* (New York: Hill and Wang, 1982); Perry Miller, "Nature and the National Ego," in *Errand into the Wilderness* (Cambridge, MA: Harvard University Press, 1984) 209; William H. Truettner, *The West as America: Reinterpreting Images of the Frontier, 1820–1920* (Washington, DC: Smithsonian Institution Press, 1991), 30; David S. Reynolds, *Walt Whitman's America: A Cultural Biography* (New York: Vintage Books, 1996); David Nye, *America as Second Creation: Technology and Narratives of New Beginnings* (Cambridge, MA: MIT Press, 2003), esp. chap. 1, "Narrating the Assimilation of Nature," 9–20; Scott A. Sandage, *Born Losers: A History of Failure in America* (Cambridge, MA: Harvard University Press, 2005); Daniel Walker Howe, *What Hath God Wrought: The Transformation of America, 1815–1841* (New York: Oxford University Press, 2007); Heather Cox Richardson, *West from Appomattox: The Reconstruction of America after the Civil War* (New Haven, CT: Yale University Press, 2007); and Kevin Rozario, *The Culture of Calamity: Disaster and the Making of Modern America* (Chicago: University of Chicago Press, 2007).

9   Ralph Waldo Emerson, *Nature*, RWE, 3, 7–8, 26, 39.

10 Emerson, *Nature*, 39.

11 Michael Williams, *Americans & Their Forests: A Historical Geography* (Cambridge: Cambridge University Press, 1989), 3–4.

12 The primary source base attesting to the strangeness of the treeless regions of the United States is enormous. Representative examples include the Long Expedition's *Account of an Expedition from Pittsburgh to the Rocky Moun-*

tains: Performed in the Years 1819 and 1820, by Order of the Hon. J. C. Calhoun, Secretary of War, under the Command of Maj. S. H. Long, of the U.S. Top. Engineers, 3 vols. (London: Longman, Hurst, Rees, Orme, and Brown, 1823), after the Lewis and Clark Expedition, the earliest well-publicized exploration of the US interior, as well as from hundreds of newspapers culled from every corner of the United States, and from every decade after the publication of the Long Expedition's findings. Quotation from "The Great American Desert," *Arkansas State Gazette*, Feb. 28, 1837.

13 For a much richer history and historiography of trees as narrative devices in the nineteenth-century United States, see Daegan Miller, "Reading Tree in Nature's Nation: Toward a Field Guide to Sylvan Literacy in the Nineteenth-Century United States," *AHR* 121, no. 4 (Oct. 2013): 1114–40, from which this paragraph was adapted.

14 In addition to works already cited, my thinking on landscape is indebted to D. W. Meinig, ed., *The Interpretation of Ordinary Landscapes: Geographical Essays* (New York: Oxford University Press, 1979); Denis Cosgrove and Stephen Daniels, eds., *The Iconography of Landscape: Essays on the Symbolic Representation, Design, and Use of Past Environments* (New York: Cambridge University Press, 1988); James Duncan and David Ley, eds., *Place/Culture/Representation* (London: Routledge, 1993); J. B. Jackson, *Landscapes: Selected Writings of J. B. Jackson*, ed. Ervin H. Zube (Amherst: University of Massachusetts Press, 1970) and John Brinkerhoff Jackson, *A Sense of Place, a Sense of Time* (New Haven, CT: Yale University Press, 1994); Paul Groth and Todd W. Bressi, eds., *Understanding Ordinary Landscapes* (New Haven, CT: Yale University Press, 1997); Matthew Potteiger and Jamie Purinton, *Landscape Narratives: Design Practices for Telling Stories* (New York: John Wiley & Sons, 1998); and Chris Wilson and Paul Groth, eds., *Everyday America: Cultural Landscape Studies after J. B. Jackson* (Berkeley: University of California Press, 2003).

15 Henry David Thoreau, *Walden; or, Life in the Woods*, WWMC, 387.

16 *Lessons from an Apple Tree: A Gift from the Teachers and Children of the Warren Street Chapel, Boston* (Boston: Dutton and Wentworth's Print, 1848), 5–8.

17 John L. O'Sullivan, editor of *The United States Magazine, and Democratic Review*, is usually given credit for coining the term *Manifest Destiny* in 1845, in his essay "Annexation," but it was most likely one of his employees, Jane McManus Storm (also known as Cora Montgomery), who came up with the phrase and first used it in an article entitled "The Great Nation of Futurity," published in 1839. See Amy Greenberg, *Manifest Manhood and the Antebellum American Empire* (New York: Cambridge University Press, 2005), 20; "The Great Nation of Futurity," *USM* 6, no. 23 (Nov. 1839): 426; "Annexation," *USM* 17, no. 85 (July–Aug. 1845): 5–10.

18 The "Second Middle Passage" comes from Ira Berlin's *Generations of Captiv-*

ity: A History of African-American Slaves* (Cambridge, MA: Harvard University Press, 2003), 161.

19   Steven Stoll, *Larding the Lean Earth: Soil and Society in Nineteenth-Century America* (New York: Hill and Wang, 2002).

20   Michael Williams, *Americans & Their Forests: A Historical Geography* (Cambridge: Cambridge University Press, 1989), 393–94; Thomas R. Cox, *The Lumberman's Frontier: Three Centuries of Land Use, Society, and Change in America's Forests* (Corvallis: Oregon State University Press, 2010), esp. chap. 8, "Actions and Reactions," 191–212; and Thomas R. Cox, Robert S. Maxwell, Phillip Drennon Thomas, and Joseph J. Malone, *This Well-Wooded Land: Americans and Their Forests from Colonial Times to the Present* (Lincoln: University of Nebraska, 1985), esp. pt. 3, "The Great Transition, 1850–1900," 111–88.

21   The discussion of modernity, capitalism, progress, and their unintended consequence is founded upon the work of the late Marshall Berman, *All That Is Solid Melts into Air: The Experience of Modernity* (New York: Penguin Books, 1982); Raymond Williams, including *The Country and the City* (New York: Oxford University Press, 1977), *Culture and Society: 1780–1950* (New York: Columbia University Press, 1983), and *Politics of Modernism: Against the New Conformists* (London: Verso, 1989); Carolyn Merchant, *The Death of Nature: Women, Ecology and the Scientific Revolution* (New York: HarperSanFrancisco, 1983); David Harvey, *The Condition of Postmodernity: An Inquiry into the Origins of Cultural Change* (Cambridge, UK: Blackwell, 1990); Anthony Giddens, *The Consequences of Modernity* (Stanford, CA: Stanford University Press, 1990); Bruno Latour, *We Have Never Been Modern*, trans. Catherine Porter (Cambridge, MA: Harvard University Press, 1993); Val Plumwood, *Feminism and the Mastery of Nature* (London: Routledge, 1993); Hannah Arendt, *The Human Condition*, 2nd ed. (Chicago: University of Chicago Press, 1998); Max Horkheimer and Theodor W. Adorno, *Dialectic of Enlightenment: Philosophical Fragments*, ed. Gunzelin Schmid Noerr, trans. Edmund Jephcott (Stanford, CA: Stanford University Press, 2002); Michel Foucault, "What Is Enlightenment?" in *The Essential Foucault: Selections from Essential Works of Foucault, 1954–1984*, ed. Paul Rabinow and Nikola Rose (New York: New Press, 2003); and Dipesh Chakrabarty, *Provincializing Europe: Postcolonial Thought and Historical Difference* (Princeton, NJ: Princeton University Press, 2008). On violence, modernity, and environmental destruction, see especially Derrick Jensen, *A Language Older Than Words* (White River Junction, VT: Chelsea Green Publishing, 2004).

22   "The legacy of domination" is a phrase I borrow from the social ecologist Murray Bookchin's *The Ecology of Freedom: The Emergence and Dissolution of Hierarchy* (Oakland, CA: AK Press, 2005).

23   Horkheimer and Adorno, *Dialectic of Enlightenment*, 1.

24   Ralph Waldo Emerson, "Ode, Inscribed to W. H. Channing," *RWE*, 693–95;

Robert D. Richardson Jr., *Emerson: The Mind on Fire* (Berkeley: University of California Press, 1995), 431–32.

25 American history is usually told in two parts: a premodern antebellum period, filled with possibility, and the post–Civil War emergence of a "modern" America, in which modernity is chiefly characterized by the cultural triumph of capitalism. But America, as Hannah Arendt points out, has always been modern. For alternative periodization of American history, see especially Lewis Perry, *Boats against the Current: American Culture between Revolution and Modernity, 1820–1860* (Lanham, MD: Rowman & Littlefield Publishers, 1993); Christopher Hager and Cody Mars, "Against 1865: Reperiodizing the Nineteenth Century," *J19* 1, no. 2 (Fall 2013): 259–84; Arendt, *Human Condition*, 248–57.

26 The key thinker on hegemony in American culture is T. J. Jackson Lears. Though Lears's use of Gramscian hegemony works well for the upper-class cultural elite that he focuses on in *No Place of Grace*, hegemony poses the temptation and the risk of seeming to be a nearly-impossible-to-escape shroud that has descended upon us all. Instead of seeing capitalistic culture as hegemonic, as "an unintended consequence of sincere (though often self-deceiving) efforts to impose moral meaning on a rapidly changing social world," as Lears put it, I'd like to take those sincere critics on their own terms and envision the culture of capitalism as always in danger of being overturned. James C. Scott is useful here: "The problem with the hegemonic thesis, at least in its strong forms as proposed by some of Gramsci's successors, is that it is difficult to explain how social change could ever originate from below." What a Marxist cultural hegemony gives us is a powerful lens for understanding why things remain the same, whereas Scott's, and other anarchist-inspired approaches, helps us understand how things change. See T. J. Jackson Lears, "The Concept of Cultural Hegemony: Problems and Possibilities," *AHR* 90, no. 3 (June 1985): 567–93; Lears, *No Place of Grace*, 5–7, 10; James C. Scott, *Domination and the Arts of Resistance: Hidden Transcripts* (New Haven, CT: Yale University Press, 1990), 78, 77–90.

27 George Fitzhugh, *Sociology for the South; or, The Failure of Free Society* (Richmond, VA: A. Morris, 1854).

28 My thinking on countermodernity was first influenced by Foucault's offhanded remark that he would like to "try to find out how the attitude of modernity, ever since its formation, has found itself struggling with attitudes of 'countermodernity'" (actually, punk rock was the first influence, but Foucault was the first genteel source). The countermoderns that I discuss are part of what Murray Bookchin has called "a legacy of freedom": "The legacy of freedom . . . lives in the daydreams of humanity in the great ideals and movements—rebellious, anarchic, Dionysian—that have welled up in all great eras of social transition." Another way to put this all is to say that the legacy of freedom depends on what Hannah Arendt called "natality," or "the

miracle that saves the world ... the birth of new men and the new beginning, the action they are capable of by virtue of being born." In addition to the theorists of modernity noted previously, my thinking on a multistranded countermodernity derives from Michel de Certeau, *The Practice of Everyday Life*, trans. Steven Rendall (Berkeley: University of California Press, 1988); Homi K. Bhabha, *The Location of Culture* (New York: Routledge Classics, 1994); James C. Scott, *Seeing Like a State: How Certain Schemes to Improve the Human Condition Have Failed* (New Haven, CT: Yale University Press, 1998), esp. "Part 4: The Missing Link," 309–57; David Harvey, *Spaces of Hope* (Berkeley: University of California Press, 2000); Jane Bennett, *The Enchantment of Modern Life: Attachments, Crossings, and Ethics* (Princeton, NJ: Princeton University Press, 2001). Quotations from Foucault, "What Is Enlightenment?" 48; Bookchin, *Ecology of Freedom*, 108; Arendt, *Human Condition*, 247. And for a beautiful riff on Arendt's natality, see Robert Pogue Harrison, *Juvenescence: A Cultural History of Our Age* (Chicago: University of Chicago Press, 2014), esp. chap. 4, "Amor Mundi," 112–44.

29 *Oxford English Dictionary* online, s.v. "radical," http://www.oed.com, accessed Apr. 16, 2015.

30 Henri Lefebvre and David Harvey have been key in helping me find countermodern nineteenth-century landscapes. Lefebvre's observations about social space as a product, his injunction that we must focus on the always-occurring process of its creation, and his argument that "a revolution that does not produce a new social space has not realized its full potential; indeed it has failed in that it has not changed life itself, but has merely changed ideological superstructures, institutions or political apparatuses," provides much of the backbone for this book. Likewise, Harvey's contention that the world is literally aboil with anticapitalist sentiment, and that spaces of hope abound if only we know how to see them, informs much of my work. ("Ask first: where is anti-capitalist struggle to be found? The answer is everywhere."). Henri Lefebvre, *Production of Space*, trans. Donald Nicholson-Smith (Malden, MA: Blackwell Publishing, 1991), 54, 30–65; Harvey, *Spaces of Hope*, 71.

31 Emily Dickinson, "41," in *The Complete Poems of Emily Dickinson*, ed. Thomas H. Johnson (Boston: Little, Brown and Company, 1960), 24.

32 The intellectual and cultural histories that have most influenced my alternative reading of the nineteenth century include Lee Clark Mitchell, *Witnesses to a Vanishing America: The Nineteenth-Century Response* (Princeton, NJ: Princeton University Press, 1981); John L. Thomas, *Alternative America: Henry George, Edward Bellamy, Henry Demarest Lloyd and the Adversary Tradition* (Cambridge, MA: Harvard University Press, 1983); Leo Marx, "Does Improved Technology Mean Progress?" *Technology Review* 90 (Jan. 1987): 33–41, 71; Casey Nelson Blake, *Beloved Community: The Cultural Criticism of Randolph Bourne, Van Wyck Brooks, Waldo Frank, & Lewis*

Mumford (Chapel Hill: University of North Carolina Press, 1990); Lawrence Buell, *The Environmental Imagination: Thoreau, Nature Writing, and the Foundation of American Culture* (Cambridge, MA: Harvard University Press, 1995); Rebecca Solnit, *Savage Dreams: A Journey into the Landscape Wars of the American West* (Berkeley: University of California Press, 1999); Sankar Muthu, *Enlightenment against Empire* (Princeton, NJ: Princeton University Press, 2003); Stephanie LeMenager, *Manifest and Other Destinies: Territorial Fictions of the Nineteenth-Century United States* (Lincoln: University of Nebraska Press, 2004); and Aaron Sachs, *The Humboldt Current: Nineteenth-Century Exploration and the Roots of American Environmentalism* (New York: Viking, 2006) and *Arcadian America: The Death and Life of an Environmental Tradition* (New Haven, CT: Yale University Press, 2013).

33 Elizabeth Kolbert, *The Sixth Extinction: An Unnatural History* (New York: Henry Holt & Co., 2014).

34 The rest of what you'll read is my attempt to follow Susan Sontag's declaration that, in addition to a hermeneutic approach to culture (she was writing of art, but it's fruitful to extend her thinking), we need an erotic one. See Susan Sontag, *Against Interpretation and Other Essays* (New York: Picador, 1966) 14. For more recent explorations of similar territory, see Eve Kosofsky Sedgwick, "Paranoid Reading and Reparative Reading, or, You're So Paranoid You Probably Think This Essay Is about You," in *Touching Affect* (Durham, NC: Duke University Press, 2003), 123–51; and Rita Felski, *The Limits of Critique* (Chicago: University of Chicago Press, 2015).

ACT ONE

1 Archibald Macleish, *Land of the Free* (New York: Harcourt, Brace and Company, 1938), 80.

2 Information on the river's current comes from Henry David Thoreau, "Plan of Concord River from East Sudbury to Billerica Mills, 22.15 Miles, To be used on a trial in the S.J. Court, Sudbury & East Sudbury Meadow Corporation vs. Middlesex Canal, Taken by agreement of Parties, by L. Baldwin, Civil Engineer. Surveyed & Drawn by B.F. Perham, May 1834," n.d. [1860?], HDTS.

3 Henry David Thoreau, *JHDT*, 1:268; 2:1464. Music is a constant theme in Thoreau's *Journals*, and one gets the sense that Thoreau was treated to the wind's symphony—one free to anybody who cares to listen—nearly everywhere he went, in every weather, and in every season. A quick note on Thoreau's *Journals*: I've consulted two different versions for this project: the Torrey and Allen 1962 reprint edition, which I cite in this note, as well as Thoreau's manuscript journal, located at the Morgan Library and Museum (here abbreviated JAM).

4   Although it is not possible to say exactly when Thoreau chopped his mark into the willow tree, it was certainly there by July 4, as he mentions it explicitly in Henry David Thoreau, "July 4th '59," 1859, MS, HDTP, vault A35, Thoreau, unit 1, box 1, folder 6. The first mention of the willow in his journal appears on the inside cover of Henry David Thoreau, JAM.

5   Thoreau has come to be canonized in a variety of ways: Leo Marx gives us a complex, solipsistic Thoreau, the pastoral writer. Roderick Frazier Nash and Max Oelschlaeger gave us a heroic Thoreau, the "philosopher of wilderness." William Cronon and Michael Pollan defrocked the heroic philosopher in the 1980s and 1990s, and instead gave us Thoreau, the dangerous ideologue. Then there is Thoreau the subversive scientist and protoecologist searching for interconnection pioneered by Donald Worster, Daniel Botkin, and Laura Dassow Walls. See Leo Marx, *The Machine in the Garden: Technology and the Pastoral Ideal in America* (London: Oxford University Press, 1976), 244–65; Roderick Frazier Nash, *Wilderness and the American Mind*, 4th ed. (New Haven, CT: Yale University Press, 2001), 84–95; Max Oelschlaeger, "Henry David Thoreau: Philosopher of the Wilderness," in *The Idea of Wilderness: From Prehistory to the Age of Ecology* (New Haven, CT: Yale University Press, 1991), 133–71; William Cronon, *Changes in the Land: Indians, Colonists, and the Ecology of New England*, 20th anniversary ed. (New York: Hill and Wang, 2003), esp. chap. 1, "The View from Walden"; and Cronon, "The Trouble with Wilderness; or Getting Back to the Wrong Nature," in *Uncommon Ground: Rethinking the Human Place in Nature*, ed. William Cronon (New York: W. W. Norton & Co., 1996), 71, 74–75; Michael Pollan, *Second Nature: A Gardener's Education* (New York: Delta Trade Paperbacks, 1991); Donald Worster, *Nature's Economy: A History of Ecological Ideas*, 2nd ed. (Cambridge: Cambridge University Press, 1994), 59–76; Daniel B. Botkin, *No Man's Garden: Thoreau and a New Vision for Civilization and Nature* (Washington, DC: Island Press, 2001); and Laura Dassow Walls, *Seeing New Worlds: Henry David Thoreau and Nineteenth-Century Natural Science* (Madison: University of Wisconsin Press, 1995).

6   Until very recently Thoreau's surveying activities were considered quaint at best, and at worst a distraction from his real work—writing—if they were acknowledged at all. For reevaluations of Thoreau's surveying work, see Albert F. McLean Jr., "Thoreau's True Meridian: Natural Fact and Metaphor," *AQ* 20, no. 3 (Autumn, 1968): 567–79; Rick Van Noy, *Surveying the Interior: Literary Cartographers and the Sense of Place* (Reno: University of Nevada Press, 2003), esp. chap. 2, "Surveying the Strange: Henry David Thoreau's Intelligence of Place," 38–72; Leslie Perrin Wilson (curator at the Concord Free Public Library and an incredibly generous scholar of all things Concord), "Thoreau's Manuscript Surveys: Getting Beyond the Surface," *CS*, n.s., 15 (2007): 24–35; John Hessler, "From Ortelius to Champlain: The Lost Maps of Henry David Thoreau," *CS*, n.s., 18 (2010): 1–26; Sarah Luria,

"Thoreau's Geopoetics," in *Geohumanities: Art, History, Text at the Edge of Place*, Michael Dear, Jim Ketchum, Sarah Luria, and Douglas Richardson, eds. (New York: Routledge, 2011). Patrick Chura is the dean of Thoreau the Surveyor. See his "Economic and Environmental Perspectives in the Surveying 'Field Notes' of Henry David Thoreau," *CS*, n.s., 15 (2007): 36–63, and *Thoreau the Land Surveyor* (Gainesville: University Press of Florida, 2010), which is the benchmark study of Thoreau's surveying practice.

7 Brian Donahue, *The Great Meadow: Farmers and the Land in Colonial New England* (New Haven, CT: Yale University Press, 2004), xv, 99–100, 155, passim.

8 For a detailed investigation of the long history of conflict over the Merrimack River watershed, see Theodore Steinberg, *Nature Incorporated: Industrialization and the Waters of New England* (Cambridge: Cambridge University Press, 1991). For a history of the struggle over the Concord River, see Brian Donahue, "'Dammed at Both Ends and Cursed in the Middle': The 'Flowage' of the Concord River Meadows, 1798–1862," *ER* 13, nos. 3–4 (Fall–Winter 1989): 47–68. The most detailed accounting of the history of the debate over Concord's water, and of Thoreau's data collection, is Robert M. Thorson, *The Boatman: Henry David Thoreau's River Years* (Cambridge, MA: Harvard University Press, 2017).

9 For the Concord Farmers' Club, see the records of the Concord Farmers' Club, held by the William Monroe Special Collections, Concord Free Public Library.

10 See Leo Marx, "Does Improved Technology Mean Progress?" *Technology Review* 90 (Jan. 1987): 34, 71; Steven Stoll, *Larding the Lean Earth: Soil and Society in Nineteenth-Century America* (New York: Hill and Wang, 2002); Philip J. Pauly, *Fruits and Plains: The Horticultural Transformation of America* (Cambridge, MA: Harvard University Press, 2007); David E. Nye, *America as Second Creation: Technology and Narratives of New Beginnings* (Cambridge, MA: MIT Press, 2003), 4, 11, 14, passim; and Scott A. Sandage, *Born Losers: A History of Failure in America* (Cambridge, MA: Harvard University Press, 2005).

11 Donahue, "'Dammed at Both Ends,'" 54–55.

12 Simon Brown, ed., *The New England Farmer; A Monthly Journal, Devoted to Agriculture, Horticulture, and Their Kindred Arts and Sciences; and Illustrated with Numerous Beautiful Engravings* (1859), 11:207.

13 Donahue, *Great Meadow*, esp. chap. 9, "Epilogue: Beyond the Meadows," 221–34.

14 I should note that not every instance of the grid runs perfectly east-west (for instance, much of Manhattan, with the exception of parts of Tribeca, SoHo, the West Village, Little Italy, and the Lower East Side, runs from the southeast to the northwest, to more perfectly square the island).

15 Penn's commitment to city building leads John Reps, one of the grid's

staunchest enemies, to conclude that "in the breadth of his outlook and his attention to detail Penn was unsurpassed as a colonial regional planner and administrator." John W. Reps *The Making of Urban America: A History of City Planning in the United States* (Princeton, NJ: Princeton University Press, 1965), 157–65. John R. Stilgoe fixes the date of the grid's American birth as 1681, but he focuses on when Penn actually gave the orders for laying out the city rather than building in the time during which the idea gestated in Penn's mind. See John Stilgoe, *Common Landscape of America, 1580–1845* (New Haven, CT: Yale University Press, 1982), 88.

16 Although we like to think that Daniel Boone always migrated west in search of solitude, he was a failed surveyor and land speculator, and it seems that he moved west continually in search of clear title to a piece of land. See John Mack Faragher, *Daniel Boone: The Life and Legend of an American Pioneer* (New York: Henry Holt and Company, 1992), 240–49.

17 Andro Linklater points out that one of the democratic aspects of the quadralinear grid was that it was easy for anyone with a basic knowledge of how a compass worked to pace out his boundaries. See his *Measuring America: How the United States Was Shaped by the Greatest Land Sale in History* (New York: Plume, 2002), 169.

18 Thomas Jefferson, *Notes on the State of Virginia* in *Thomas Jefferson: Writings* (New York: Library of America, 1984), 290–91.

19 See J. B. Jackson, "Jefferson, Thoreau & After," in *Landscapes: Selected Writings of J. B. Jackson*, ed. Ervine H. Zube (Amherst: University of Massachusetts Press, 1970), 4, 5. For the equation of the grid with providential design, see especially Nye, *America as Second Creation*, chap. 2, "Surveying the Ground," 21–42.

20 It was Melville who would pen one of the most enduring lines for all who love to wander: "It is not down in any map; true places never are." Herman Melville, *Moby Dick* (1851; New York: Barnes & Noble Classics, 2003), 27, 85.

21 E. O. Wilson, introduction to *From So Simple a Beginning: The Four Great Books of Charles Darwin*, ed. E. O. Wilson (New York: W. W. Norton & Co., 2006), 17. For more on the midcentury romantic surveys, see William H. Goetzmann, *Exploration and Empire: The Explorer and the Scientist in the Winning of the American West* (New York: Monticello Editions, 1966); Wallace Stegner, *Beyond the Hundredth Meridian: John Wesley Powell and the Second Opening of the West* (New York: Penguin Books, 1992); Aaron Sachs, *The Humboldt Current: Nineteenth-Century Exploration and the Roots of American Environmentalism* (New York: Viking, 2006).

22 And in a nicely reciprocal way, Edgar Allan Poe's *The Narrative of Arthur Gordon Pym of Nantucket* is based in part on the report of J. N. Reynolds's 1836 *Address on the Subject of a Surveying and Exploring Expedition to the Pacific Ocean and South Seas*, given before Congress in support of exploration.

23 Thoreau was even in personal communication with one of the nation's foremost astronomers, William Cranch Bond, the head of the Harvard Observatory, on the subject of terrestrial magnetism. Henry D. Thoreau, "'Lovering and Bond on Mag. Observation at Cambridge,' Am. Acad. 1846." n.d., MS, HDTP 1836–1862, vault A35, Thoreau, unit 1, box 1, folder 9; William C. Bond to Thoreau, "Diurnal Magnetic variation . . ." June 9, 1851, MS, HDTP, vault A35, Thoreau, unit 1, box 1, folder 9; Henry David Thoreau, "[Scene at?] the Cam. Observatory and Library," n.d., MS, HDTP, vault A35, Thoreau, unit 1, box 1, folder 9.

24 Patrick Chura argues that Thoreau first learned to survey when he and his brother John taught the subject in the school they ran, sometime between 1839 and 1841, and Thoreau's chief biographer, Walter Harding, notes that in 1840 Thoreau purchased two surveying instruments for use in the school he and his brother John had opened two years earlier. See Chura, "Economic and Environmental Perspectives," 38; Walter Harding, *The Days of Henry Thoreau: A Biography* (Princeton, NJ: Princeton University Press, 1982), 83–84.

25 Harding, *Days of Henry Thoreau*, 235; Chura, *Thoreau the Land Surveyor*, 71.

26 Charles Davies, LL.D, *Elements of Surveying, and Navigation; With a Description of the Instruments and the Necessary Tables*, rev. ed. (New York: A. S. Barnes & Co., 1847), in William Monroe Special Collections, Concord Free Public Library. On July 3, 1851, for instance, Thoreau checked his compass nine times between 7 a.m. and 7 p.m. Furthermore, Thoreau performed the difficult experiment of comparing his compass to the North Star at least two times, once in 1840, and once, to his apparent joy, in 1851: "Found the direction of the Pole Star with greatest *Meridian Elongation* . . . at 9h26m PM!" Henry D. Thoreau, *FN*, 38, 63; and "Bearing By the Compass—which varied 9° West of the Pole," HDTP, vault A35, Thoreau, unit 1, box 1, folder 9.

27 Henry David Thoreau, "Walking," in *CEP*, 234. Chura is especially good on exploring the literary effect of Thoreau's surveying. See Chura, *Thoreau the Land Surveyor*, esp. chap. 2, "Material to Mythology," 22–44.

28 Only a fragment of the letter hiring Thoreau survives. See Simon Brown, David Heard, John W. Simonds, and Samuel H. Rhoades to Thoreau, June 4, 1859, HDTP, vault A35, Thoreau, unit 1, box 1, folder 6. The Joint Special Committee issued a fascinating report a year later. See *Report of the Joint Special Committee upon the Subject of the Flowage of Meadows on Concord and Sudbury Rivers* (Boston: William White, Printer to the State, 1860).

29 Committee of the Proprietors of Sudbury and Concord River Meadows, River Meadow Committee for the Town of. . . 1859, HDTP, vault A35, Thoreau, unit 1, box 1, folder 6.

30 Thoreau, JAM, July 5 and July 12.

31 Brown to Thoreau, June 4, 1859; Henry David Thoreau, *CHDT*, 522; David Heard to Thoreau, July 1, 1859, MS, HDTP, vault A35, Thoreau, unit 1, box 1,

folder 7; Jon. A. Hill to Thoreau, July 25, 1859, MS, HDTP, vault A35, Thoreau, unit 1, box 1, folder 7; C. C. Thackford to Thoreau, n.d. [1859?], HDTP, vault A35, Thoreau, unit 1, box 1, folder 7; Henry David Thoreau, Miscellaneous notes from river survey and statistics on the back of an envelope addressed to "Henry D. Thoreaux" [1858–1859?], MS, HDTP, vault A35, unit 1, box 1, folder 6; Thoreau, JAM, June 24 and July 14.

32 Henry David Thoreau, Notes on Bridges, n.d. [1858–1859?], MS, HDTP, vault A35, Thoreau, unit 1, box 1, folder 6; Henry David Thoreau, Draft of Statistics of the Bridges over Concord River, 1859, MS, HDTP, vault A35, Thoreau, unit 1, box 1, folder 5. For the final version, see Henry David Thoreau, *Statistics of the Bridges over Concord River, between Heard's Bridge and Billerica Dam, Obtained June 22nd, 23rd, & 24th 1859; The Level of the Water at Concord in the Meantime, not having Varied One Inch from about 3 Feet Above Summer Level*, 1859, HDTS.

33 Ralph Waldo Emerson to Elizabeth Hoar, Aug. 3, 1859, Morgan Library. After Thoreau's death, Emerson was asked to write a biographical sketch for an edition of Thoreau's writings called *Excursions*. Perhaps feeling that he had spoken ungenerously toward Thoreau on too many occasions — including, famously, during his eulogy for his deceased erstwhile student — Emerson used the occasion to cast a different, kinder view of Thoreau's river survey: "Mr. Thoreau dedicated his genius with such entire love to the fields, hills, and waters of his native town, that he made them known and interesting to all reading Americans, and to people over the sea. The river on whose banks he was born and died he knew from its springs to its confluence with the Merrimack. He had made summer and winter observations on it for many years, and at a every hour of the day and the night. The result of the recent survey of the Water Commissioners appointed by the State of Massachusetts he had reached by his private experiments, several years earlier. Every fact which occurs in the bed, on the banks, or in the air over it . . . were all known to him, and, as it were, townsmen and fellow-creatures; so that he felt an absurdity or violence in any narrative of one of these by itself apart, and still more of its dimensions on an inch-rule, or in the exhibition of its skeleton, or the specimen of a squirrel or a bird in brandy. He liked to speak of the manners of the river, as itself a lawful creature, yet with exactness, and always to an observed fact." See Emerson, "Thoreau," in *RWE*, 816.

34 Lewis M. Haupt, *The Topographer, His Instruments and Methods. Designed for the use of Students, Amateur Topographers, Surveyors, Engineers, and All Persons Interested in the Location and Construction of Works Based Upon Topography. Illustrated with Numerous Plates, Maps, and Engravings* (New York: J. M. Stoddart, 1883), xiii. Exposing the larger cultural dimensions underlying scientific practice has long been one of the mainstays of science and technology studies and history of science. Relevant works include Nathan Reingold, *Science, American Style* (New Brunswick, NJ: Rutgers University

Press, 1991); Lorraine Daston and Peter Galison, "The Image of Objectivity," *Representations* 40 (1992): 81–128; Robert E. Kohler, *Landscapes and Labscapes: Exploring the Lab-Field Border in Biology* (Chicago: University of Chicago Press, 2007); and Steven Shapin's magnificent collection of essays, *Never Pure: Historical Studies of Science as If It Was Produced by People with Bodies, Situated in Time, Space, Culture, and Society, and Struggling for Credibility and Authority* (Baltimore: Johns Hopkins University Press, 2010).

35 This is one of the key insights of critical, poststructural readings of cartography, and extends beyond American history. See, for example, Thongchai Winichakul, *Siam Mapped: A History of the Geo-Body of a Nation* (Honolulu: University of Hawai'i Press, 1994); Matthew Edney, *Mapping an Empire: The Geographical Construction of British India, 1765–1843* (Chicago: University of Chicago Press, 1997); D. Graham Burnett, *Masters of All They Surveyed: Exploration, Geography, and a British El Dorado* (Chicago: University of Chicago Press, 2000); J. B. Harley, *The New Nature of Maps: Essays in the History of Cartography*, ed. Paul Laxton (Baltimore: Johns Hopkins University Press, 2001), and Raymond Craib's excellent, *Cartographic Mexico: A History of State Fixations and Fugitive Landscapes* (Durham, NC: Duke University Press, 2004).

36 Horkheimer and Adorno point out that abstraction paved the way for the monetization of everything—nature, human bodies, and culture. See Max Horkheimer and Theodor W. Adorno, *Dialectic of Enlightenment: Philosophical Fragments*, ed. Gunzelin Schmid Noerr, trans. Edmund Jephcott (Stanford, CA: Stanford University Press, 2002), 9, 12, passim. My thinking on the commodity is rooted more generally in Karl Marx, *Capital*, vol. 1 (1867; repr., London: Penguin Classics, 1990), esp. "Part One: Commodities and Money," 125–244.

37 I'm thinking here with J. B. Jackson's classic distinction between political and inhabited landscapes. John Brinckerhoff Jackson, "A Pair of Ideal Landscapes," in *Discovering the Vernacular Landscape* (New Haven, CT: Yale University Press, 1984), 11, 14, 15, 28, 42, 54.

38 Aaron Sachs, *Arcadian America: The Death and Life of an Environmental Tradition* (New Haven, CT: Yale University Press, 2013); Reps, *Making of Urban America*, 325–36; Shen Hou, *The City Natural: Garden and Forest Magazine and the Rise of American Environmentalism* (Pittsburgh, PA: University of Pittsburgh Press, 2013).

39 This intuition has become foundational among many scholars who work on land, space, and American culture. For influential examples, see Henry Nash Smith, *Virgin Land: The American West as Symbol and Myth* (Cambridge, MA: Harvard University Press, 1978), 12; Leo Marx, "The American Ideology of Space," in *Denatured Visions: Landscape and Culture in the Twentieth Century*, ed. Stuart Wrede and William Howard Adams (New York: Museum of Modern Art, New York, 1991), 63; Nye, *America as Second Creation*, 25;

Karl Jacoby, *Crimes against Nature: Squatters, Poachers, Thieves and the Hidden History of American Conservation* (Berkeley: University of California Press, 2001), 29.

40 William Cronon, *Nature's Metropolis: Chicago and the Great West* (New York: W. W. Norton & Co., 1991), 102, 114–23; Boutmy quoted on 53–55.
41 Linklater, *Measuring America*, 149–50.
42 Harding, *Days of Henry Thoreau*, 276–77.
43 Thoreau, "Field Notes," 69; Thoreau, *JHDT*, 1:270.
44 "What the map cuts up, the story cuts across," writes Michel de Certeau. "Stories thus carry out a labor that constantly transforms places into spaces or spaces into places. They also organize the play of changing relationships between places and spaces." Michel de Certeau, *The Practice of Everyday Life*, trans. Steven Rendall (Berkeley: University of California Press, 1988), 117, 129.
45 Thoreau, *JHDT*, 1:247.
46 Thoreau, *JHDT*, 2:1476. Thoreau's surveying journal ends sometime in February 1858, and from then on, he seemed to keep his surveying notes along with his personal journal.
47 Thoreau, *JHDT*, 2:1482.
48 Thoreau, *JHDT*, 1:579.
49 For the question of Thoreau's manifold relationship to something we might call "modern," see François Specq, Laura Dassow Walls, and Michael Granger, eds., *Thoreauvian Modernities: Transatlantic Conversations on an American Icon* (Athens: University of Georgia Press, 2013); and Jerome Tharaud, "'So far Heathen': Thoreau, The Missionary Memoir, and Walden's Cosmic Modernity," *ESQ* 59, no. 4 (2013): 618–61.
50 They also probably knew of Thoreau's work as an expert witness frequently called to give testimony against dams, which he did in 1851, 1853, and 1854. Thoreau, *JHDT*, 1:211; Harding, *Days of Henry Thoreau*, 326; Kenneth Walter Cameron, "Thoreau in the Court of Common Pleas (1854)," *Emerson Society Quarterly* 14, no. 1 (1959): 86–89.
51 See Donahue, "'Dammed at Both Ends.'"
52 Thoreau, *JHDT*, 1:33.
53 Thoreau, *JHDT*, 2:1489.
54 Thoreau waffled: in 1856 he thought that the only way the Great Meadows could have remained free of tress and bushes was if frequent floods killed the woody invaders; hence, floods that occasionally stayed too long and killed the grasses were simply to be expected; then in 1857, after consulting a 1654 history of Concord, he changed his mind. The grass-killing floods were human-made. After his research in 1859 at the behest of Brown, he was convinced: the floods were definitely artificial. But then, in 1860, he once again found himself scratching his head, unsure. Thoreau, *JHDT*, 2:1056, 1118, 1485, 1624–25, 1689.

55 Thoreau, *JHDT*, 1:789.
56 Thoreau, *JHDT*, 2:1500.
57 August 2, 1859, was typical, when he took readings of the river's depth at 6 a.m., 2 p.m., and 8 p.m. Thoreau, JAM, Aug. 2.
58 Thoreau, *JHDT*, 2:1494.
59 Thoreau, *JHDT*, 2:1504.
60 Henry David Thoreau, *W*, 37. LeMenager points out that *A Week on the Concord and Merrimack Rivers* can be read as a book contrasting two different ways of valuing a river: the commercial, and the poetic. Stephanie LeMenager, *Manifest and Other Destinies: Territorial Fictions of the Nineteenth-Century United States* (Lincoln: University of Nebraska Press, 2004), 121.
61 Thoreau, *JHDT*, 2:1011. Eric Wilson writes that Thoreau "wished to embody the pulsations of the world," and Sarah Luria makes the case for reading his river survey and *A Week on the Concord and Merrimack Rivers* alongside each other. See Wilson, *Romantic Turbulence: Chaos, Ecology and American Space* (New York: St. Martin's Press, 2000), 96; Luria, "Thoreau's Geopoetics."
62 Thoreau, *JHDT*, 2:1587.
63 Thoreau, *JHDT*, 1:425.
64 *Oxford English Dictionary* online, s.v. "abstract," http://www.oed.com, accessed Sept. 14, 2015.
65 On Frémont, the Mexican-American War, and surveying, see especially Goetzmann, *Exploration and Empire*, 240–64; for a different take on Frémont, one that captures the cultural ambivalence that his exploration helped stir up, see Anne Farrar Hyde, *An American Vision: Far Western Landscape and National Culture, 1820–1920* (New York: New York University Press, 1990), esp. chap. 1, "Looking Far West: Assessing the Possibilities of the Landscape, 1800–1850," 12–52.
66 Roger G. Kennedy, *Mr. Jefferson's Lost Cause: Land, Farmers, Slavery, and the Louisiana Purchase* (New York: Oxford University Press, 2003).
67 This paragraph leans on Edward E. Baptist's *The Half Has Never Been Told: Slavery and the Making of American Capitalism* (New York: Basic Books, 2014), 3, 114 (for statistics), and esp. chap. 3, "Right Hand, 1815–1819," and chap. 7, "Seed, 1829–1837," 215–59. For the role of cotton in the rise of global capitalism and the ascendency of the West, see Sven Beckert, *Empire of Cotton: A Global History* (New York: Alfred A. Knopf, 2015).
68 Beckert, *Empire of Cotton*, ix.
69 Baptist, *The Half Has Never Been Told*, 317.
70 Henry David Thoreau, "Slavery in Massachusetts," in *CEP*, 333; Henry David Thoreau, "Life without Principle," in *CEP*, 348–49.
71 Thoreau, "Walking," 225.
72 Henry David Thoreau, *Walden; or, Life in the Woods*, WWMC, 343–44.
73 Harding, *Days of Henry Thoreau*, 199, 202.
74 Henry David Thoreau, "Civil Disobedience," in *CEP*, 207.

75 Henry David Thoreau, "Walking," *CEP*, 240.
76 Harding, *Days of Henry Thoreau*, 200–201. For an exploration of how thoroughly the culture of reform affected American culture, especially its literary culture, see David S. Reynolds, *Beneath the American Renaissance: The Subversive Imagination in the Age of Emerson and Melville* (Cambridge, MA: Harvard University Press, 1988), esp. pt. 1, "God's Bow, Man's Arrows: Religion, Reform, and American Literature," 15–165. For an in-depth analysis of how the culture of reform made its presence felt within the ranks of transcendentalists, see Anne C. Rose, *Transcendentalism as a Social Movement, 1830–1850* (New Haven, CT: Yale University Press, 1981); and Philip F. Gura, *American Transcendentalism: A History* (New York: Hill and Wang, 2007).
77 Jonathan H. Earle does much to rehabilitate the northern faction of the Democratic Party and is an especially good source for information on Evans. Jonathan H. Earle, *Jacksonian Antislavery & the Politics of Free Soil, 1824–1854* (Chapel Hill: University of North Carolina Press, 2004), 27–37.
78 Thoreau, *Walden*, 304.
79 Stanley Cavell proposes that loss is one-half of the central theme of Thoreau's work, especially in *Walden*. Stanley Cavell, *The Senses of Walden: An Expanded Edition* (Chicago: University of Chicago Press, 1992).
80 Thoreau, *Walden*, 387. Hannah Arendt argues that one of the triumphs of the culture of modern capitalism is to reduce everything to the most basic metabolic functions—we either labor, or we consume. Such reduction—another kind of abstraction—denies the one thing that makes us fundamentally human: culture. Hannah Arendt, *The Human Condition*, 2nd ed. (Chicago: University of Chicago Press, 1998), esp. pt. 3, "Labor," 79–135.
81 The position was outlined, briefly, in Perry Miller's "Nature and the National Ego," 204–5, 207–8, 215; elaborated and celebrated in Nash's *Wilderness and the American Mind* (1967) and Oelschlaeger's *The Idea of Wilderness* (1991); and codified in Cronon's *Changes in the Land* (1983) and "The Trouble with Wilderness," (1995)—two texts that, I would argue, mark a more or less stable consensus among environmental historians as to Thoreau's intellectual legacy. But in a recent article in *Environmental History*, Kent Curtis has argued that environmental historians tend to use Thoreau as a device rather than a source. Kent Curtis, "The Virtue of Thoreau: Biography, Geography and History in Walden Woods," *EH* 15, no. 1 (Jan. 2010): 37. See also Rebecca Solnit, *Storming the Gates of Paradise: Landscapes for Politics* (Berkeley: University of California Press, 2007), 4–8, for a useful corrective.
82 See, for instance, the National Park Service's website for Death Valley National Park, at http://www.nps.gov/deva/learn/nature/wilderness.htm (accessed Sept. 27, 2015), which misquotes Thoreau as intoning "in wilderness is the preservation of the world."
83 Laura Dassow Walls, "Believing in Nature: Wilderness and Wildness in Thoreauvian Science," in *Thoreau's Sense of Place: Essays in American Envi-*

ronmental Writing, ed. Richard J. Schneider (Iowa City: University of Iowa Press, 2000) 15; Walls, *Seeing New Worlds,* 13–14, passim; Botkin, *No Man's Garden.*

84 Thoreau, "Walking," 240. For a current elaboration of Thoreau's insight, see Derrick Jensen, *A Language Older Than Words* (White River Junction, VT: Chelsea Green Publishing, 2004).

85 Thoreau, *Walden,* 501–11. For Thoreau and his role in Concord's Underground Railroad, see *Walden,* 443; *JHDT,* 1:123, 283, 646.

86 Thoreau, "Walking," 246.

87 Ibid., 225, 254.

88 Thoreau, *Walden,* 329.

89 Ibid., 587.

90 See Thoreau, "Walking," 230.

91 Hannah Arendt writes of surveying: "It is in the nature of the human surveying capacity that it can function only if man disentangles himself from all involvement in and concern with the close at hand and withdraws himself to a distance from everything near him. The greater the distance between himself and his surroundings, world or earth, the more he will be able to survey and to measure and the less will worldly, earth-bound space be left to him." Arendt, *Human Condition,* 251.

92 Thoreau, *JHDT,* 2:1240.

93 In November 1850, for instance, Thoreau consulted seven map-bearing books from the sixteenth and seventeenth centuries, including what was probably Abraham Ortelius's *Theatrum Orbis Terrarum,* or the world's first atlas. See Henry David Thoreau, "Saw at Cambridge Today . . . ," Nov. 18, 1850, HDTP, vault A35, Thoreau, unit 1, box 1, folder 9.

94 Thoreau, *JHDT,* 2:1721.

95 Eric Wilson notes that Thoreau literally sought to write a river in *A Week on the Concord and Merrimack Rivers* and a plant in *Walden.* Wilson, *Romantic Turbulence,* 95. See also Cavell's essay "Sentences" in *Senses of Walden,* 36–69.

96 On the back of Simon Brown's June letter requesting that he collect statistics on the town's bridges, Thoreau had written to himself that he needed to consult a copy of Baldwin's map, and then crossed the map through, as if, at a later point, he had checked it off his list. Brown et al. to Thoreau, June 4, 1859.

97 J. B. Harley reminds us that silences are also an active performance. See Harley, *The New Nature of Maps,* 86, 87, 97, 99. Jonathan Smith, in "The Lie That Binds: Destabilizing the Text of Landscape," in *Place/Culture/Representation,* ed. James Duncan and David Ley (London: Routledge, 1993), 78–94, takes off from Harley's argument and shows how irony also can infuse spatial representations. Irony, for Smith, is the intrusion of local, private knowledge into the blank official space of the map.

98 Then, on July 7, he wrote: "But, though meandering, it is straighter in its general course than would be believed. These nearly twenty-three miles in length (or 16+ direct) are contained within a breadth of two miles twenty-six rods; *i.e.*, so much it takes to meander in. It can be plotted by the scale of one thousand feet to an inch on a sheet of paper seven feet one and one quarter inches long by eleven inches wide." The dimensions of that sheet of paper are very nearly the dimensions of his finished map (seven feet seven inches by fifteen inches). Thoreau, *JHDT*, 2:1488.

99 Henry David Thoreau, "Names of Bridges on Baldwin's Map of 1811," [1859?], MS, HDTP, vault A35, Thoreau, unit 1, box 1, folder 6.

100 Thoreau, *JHDT*, 2:1009.

101 Cavell writes, "To realize where we are and what we are living for, the conditions of our present, the angle at which we stand to the world [notice Cavell's slide into surveying analogies], the writer calls 'improving the time,' using a preacher's phrase and giving his kind of turn to it." Cavell, *Senses of Walden*, 61. Thoreau, *Journals* 1:932.

102 "In any weather, at any hour of the day or night, I have been anxious to improve the nick of time . . . to stand on the meeting of two eternities, the past and the future, which is precisely the present moment." Thoreau, *Walden*, 336.

103 Henry David Thoreau, Map Tracings, n.d. [1857–1860?], five sheets, MS, HDTP, vault A35, Thoreau, unit 1, box 1, folder 6.

104 Thoreau, "Plan of Concord River."

105 Thoreau, "Life without Principle," 362.

106 *Report of the Joint Special Committee*, cii–ciii.

107 Thoreau, "Plan of Concord River."

108 I've borrowed the term *deep map* from William Least Heat-Moon, whereas Luria prefers the analogy "deep focus shot." In a similar vein, Laura Dassow Walls has argued that Thoreau's scientific gaze sought consilience, "the murmur of multiple voices and actions," rather than neatly separate, cleanly divisible truths. David Nye notes that many countermodern narratives take as their founding presupposition that the land is not empty, but already claimed, and if that's true, then Thoreau's map is explicitly political, a sort of prefiguring of the sort of revolutionary space imagined by Henri Lefebvre. See William Least Heat-Moon, *PrairyErth: A Deep Map* (Boston: Houghton Mifflin, 1991); Luria, "Thoreau's Geopoetics," 135; Walls, *Seeing New Worlds*, 13, 126, 132, 169, 251; Nye, *America as Second Creation*, 292, 294; and Henri Lefebvre, *The Production of Space*, trans. Donald Nicholson-Smith (Malden, MA: Blackwell Publishing, 1991).

109 Thoreau, *W*, 235.

110 Thoreau, "Huckleberries," 496.

111 Thoreau, *JHDT*, 1:539. Anne Baker notes that "Thoreau adapts the process of measurement to his own ends, transforming it into a means of resisting

the nationalist and commercial agendas that surveying served." We could also think of Thoreau's map as heterotopic, as seeking "to create a space of illusion that exposes every real space, all the sites inside of which human life is partitioned, as still more illusory," as Michel Foucault put it. See Anne Baker, *Heartless Immensity: Literature, Culture, and Geography in Antebellum America* (Ann Arbor: University of Michigan Press, 2006) 48; and Michel Foucault, "Of Other Spaces," *Diacritics* 16, no. 1 (Spring 1986): 27.

112 William Ellery Channing to Thoreau, n.d. [1859?], MS, HDTP, vault A35, Thoreau, unit 1, box 1, folder 7.

113 Thoreau, "Plan of Concord River." Luria argues—and I would agree—that this is one of the places where Thoreau denies the universal pretensions of surveying by actively writing himself—in the use of that first-person *my*—into his map's text, a bit of subjectivity as uncomfortable among scientists as it is among academics, and for similar reasons. Luria, "Thoreau's Geopoetics," 135.

114 Thoreau, "Walking," 239; *JHDT*, 2:1496.

115 Thoreau, "Walking," 230.

116 Henry David Thoreau, *The Maine Woods*, in *WWMC*, 646. This was an engrossing passion for Thoreau, and the eleventh post in his journal reads, "My desire is to know *what* I have lived, that I may know *how* to live henceforth." Thoreau, *JHDT*, 1:20.

117 Thoreau, *Walden*, 336.

118 Henry David Thoreau, *The Dispersion of Seeds*, in *FS*, 63–64.

ACT TWO

1 Joseph Bruchac, "At the End of Ridge Road: From a Nature Journal," in *The Colors of Nature: Culture, Identity, and the Natural World*, ed. Alison H. Deming and Lauret E. Savoy (Minneapolis: Milkweed Editions, 2002), 50.

2 Ralph Waldo Emerson, "The Adirondacs," in *May-Day and Other Pieces* (Boston: Houghton, Mifflin and Company, 1881) 43.

3 Henry David Thoreau, *JHDT*, 2:1344.

4 Paul Schneider, *The Adirondacks: A History of America's First Wilderness* (New York: Henry Holt and Company, 1997), 31, 87–88, 93–96, 105–106; Nathaniel Sylvester, *Historical Sketches of Northern New York and the Adirondack Wilderness* (1877; repr., Fleischmanns, NY: Harbor Hill Books, 1997), 152–65. Brown University is named for the entire Brown family.

5 Emerson, "Adirondacs," 45. For a history of Emerson's trip to Follensby Pond, see James Schlett, *A Not Too Greatly Changed Eden: The Story of the Philosophers' Camp in the Adirondacks* (Ithaca, NY: Cornell University Press, 2015).

6 Melissa Otis, "Location of Exchange: Algonquian and Iroquoian Occupation in the Adirondacks before and after Contact," *Environment, Space, Place* 5, no. 2 (2013): 7–34.
7 For more detailed histories of the Adirondacks and the park, see Alfred L. Donaldson, *A History of the Adirondacks*, 2 vols. (1921; repr., Fleischmanns, NY: Purple Mountain Press, 1996); Frank Graham Jr., *The Adirondack Park: A Political History* (Syracuse, NY: Syracuse University Press, 1978); Philip G. Terrie, *Forever Wild: A Cultural History of Wilderness in the Adirondacks* (Syracuse, NY: Syracuse University Press, 1994) and *Contested Terrain: A New History of Nature and People in the Adirondacks* (Syracuse, NY: Syracuse University Press, 1997); Caroline Mastin Welsh, ed., *Adirondack Prints and Printmakers: The Call of the Wild* (Syracuse, NY: Syracuse University Press, 1998); Georgia B. Barnhill, *Wild Impression: The Adirondacks on Paper: Prints in the Collection of the Adirondack Museum* (Boston: David R. Godine, 1995); Schneider, *Adirondacks*; Karl Jacoby, *Crimes against Nature: Squatters, Poachers, Thieves and the Hidden History of American Conservation* (Berkeley: University of California Press, 2001); Stephen B. Sulavik, *Adirondack: Of Indians and Mountains, 1535–1838* (Fleischmanns, NY: Purple Mountain Press, 2005); David Stradling, *The Nature of New York: An Environmental History of the Empire State* (Ithaca, NY: Cornell University Press, 2010).
8 Sulavik, *Adirondack*, 17, 92–93. It wasn't until 1892, when the state created the Adirondack Park and threw an official boundary line around 2.8 million acres of land, declaring much of it protected wilderness, that the Adirondacks became a defined place. Even so, that place refused to stay put: in 1912, the park was enlarged and land that was formerly "other" became, officially, Adirondack. This happened again in 1972.
9 Ebenezer Emmons, *Geology of New-York* (Albany, NY: W. & A. White & J. Vischer, 1842), pt. 2, 10–11.
10 Quoted in Sulavik, *Adirondack*, 93.
11 Sulavik, *Adirondack*, 13, 36, 38
12 Otis, "Location of Exchange," 11.
13 Charles Fenno Hoffman, one of the earliest promoters of Adirondack adventure wrote with a nostalgic sigh, "And when this remnant of the Iroquois shall have dwindled from among us, their names will still live in the majestic lakes and noble rivers that embalm the memory of their language." C. F. Hoffman, *Wild Scenes in the Forest and Prairie* (London: Richard Bentley, 1839), 2:5.
14 For the cartographic history of the Adirondacks, see Jerold Pepper, "When Men and Mountains Meet: Mapping the Adirondacks," in *Adirondack Prints and Printmakers: The Call of the Wild*, ed. Caroline Mastin Welsh (Syracuse, NY: Syracuse University Press, 1998), 1–24.
15 Ralph Waldo Emerson, *Nature*, in *RWE*, 6.

16 Emerson, "Adirondacs," 45; Emerson, *Nature*, 6.
17 Emerson, "Adirondacs," 48, 49, 55.
18 Ibid., 57.
19 Ibid., 58, 60–61.
20 Emerson, *Nature*, 6.
21 Emerson, "Adirondacs," 62.
22 The Wilderness Act of 1964, in *The Great New Wilderness Debate*, ed. J. Baird Callicott and Michael P. Nelson (Athens: University of Georgia Press, 1998), 121. The classic histories of wilderness include (in the celebratory vein) Roderick Frazier Nash, *Wilderness & the American Mind*, 4th ed. (New Haven, CT: Yale University Press, 2001); and Max Oelschlaeger, *The Idea of Wilderness: From Prehistory to the Age of Ecology* (New Haven, CT: Yale University Press, 1991); more ambivalent, complicated takes on wilderness can be found in Perry Miller, *Errand into the Wilderness* (Cambridge, MA: Harvard University Press, 1984); Leo Marx, *The Machine in the Garden: Technology and the Pastoral Ideal in America* (London: Oxford University Press, 1976); Henry Nash Smith, *Virgin Land: The American West as Symbol and Myth* (Cambridge, MA: Harvard University Press, 1978); and Lee Clark Mitchell, *Witnesses to a Vanishing America: The Nineteenth-Century Response* (Princeton, NJ: Princeton University Press, 1981).
23 William Cronon, "The Trouble with Wilderness; or, Getting Back to the Wrong Nature," in *Uncommon Ground: Rethinking the Human Place in Nature*, ed. William Cronon (New York: W. W. Norton & Co., 1995), 69–90. See also Ramachandra Guha, "Radical American Environmentalism and Wilderness Preservation: A Third World Critique," *Environmental Ethics* 11 (Spring 1989): 71–83; Barry Lopez, "Unbounded Wilderness," *Aperture* 120 (Late Summer 1990): 2–15; and the two invaluable collections that catch the emergence of the wilderness critique from the mid-1980s to the 2000s, both edited by J. Baird Callicott and Michael P. Nelson, *The Great New Wilderness Debate* (Athens: University of Georgia Press, 1998) and *The Wilderness Debate Rages On: Continuing the Great New Wilderness Debate* (Athens: University of Georgia Press, 2008).
24 See, for instance, Jacoby, *Crimes against Nature*; Louis S. Warren, *The Hunter's Game: Poachers and Conservationists in Twentieth-Century America* (New Haven, CT: Yale University Press, 1997); Mark David Spence, *Dispossessing the Wilderness: Indian Removal and the Making of the National Parks* (New York: Oxford University Press, 1999); Rebecca Solnit, *Savage Dreams: A Journey into the Landscape Wars of the American West* (Berkeley: University of California Press, 1999).
25 Cronon, "Trouble with Wilderness," 73.
26 See Schneider, *Adirondacks*, xi; Nash, *Wilderness and the American Mind*, 116; Terrie, *Forever Wild*, 4.
27 Ralph Waldo Emerson, "The Fugitive Slave Law," in *RWE*, 779, 780; and

Robert D. Richardson, *Emerson: The Mind on Fire* (Berkeley: University of California Press, 1995), 496.

28 Richardson, *Emerson*, 497–98, 545; and Walter Harding, *The Days of Henry Thoreau: A Biography* (Princeton, NJ: Princeton University Press, 1982), 415–16. Not to be outdone, Thoreau, too, would give public lectures on Brown, which were some of the most well attended events he ever gave.

29 I'm leaning heavily here, and throughout this chapter, on some of the suggestions Conevery Bolton Valenčius offers in her *The Health of the Country*—perhaps most importantly the insight that human body and landscape were understood to be intimately linked. Valenčius shows that "the geography of health" points to a "surprising holism in the worldview of the bustling, rapidly industrializing nineteenth century." Conevery Bolton Valenčius, *The Health of the Country: How American Settlers Understood Themselves and Their Land* (New York: Basic Books, 2002), 3. See also Gregg Mitman's rousing "In Search of Health: Landscape and Disease in American Environmental History," *EH* 10, no. 2 (Apr. 2005): 184–210. For a radical environmental politics of care, see Val Plumwood, *Feminism and the Mastery of Nature* (London: Routledge, 1993), 182–89.

30 I refer to the African Americans who settled in the Adirondacks as pioneers because that's what they often called themselves. See, for instance, Rev. Theodore S. Wright, Rev. Charles B. Ray, and Dr. J. M'Cune Smith, *An Address to the Three Thousand Colored Citizens of New-York, Who are the Owners of One Hundred and Twenty Thousand Acres of Land in the State of New York, Given to Them by Gerrit Smith, Esq., of Peterboro* (New York: n.p., 1846), 5.

31 It has been common practice to refer to the whole of the Adirondack experiment as Timbuctoo—I've resisted doing so to preserve the particularity of each settlement in the hopes that one day we'll have a history tracing the interactions of each interconnected microcommunity.

32 Not much has been written about this facet of Adirondack, environmental, abolitionist, and African American history, and what little there is has been written as an offshoot of John Brown. The earliest efforts include Donaldson's racist treatment in *History of the Adirondacks*, 3–12; Octavius Brooks Frothingham's *Gerrit Smith: A Biography* (New York: G. P. Putnam's Sons, 1909), 101–13; and Ralph Volney Harlow's, *Gerrit Smith: Philanthropist and Reformer* (New York: Henry Holt and Company, 1939), 242–52. For more recent treatments, see Katherine Butler Jones (who is the great-granddaughter of a pioneer), "They Called It Timbuctoo," *Orion* 17, no. 1 (Winter 1998): 27–33; Russell Banks's historical novel *Cloudsplitter: A Novel* (New York: HarperPerennial, 1999); John Stauffer, *The Black Hearts of Men: Radical Abolitionists and the Transformation of Race* (Cambridge, MA: Harvard University Press, 2001), esp. chap. 5, "Bible Politics and the Creation of the Alliance," 134–81; Milton C. Sernett, *North Star Country: Upstate New York and the Crusade for African American Freedom* (Syracuse,

NY: Syracuse University Press, 2002), esp. chap. 8, "John Brown's Body," 195–221; David S. Reynolds, *John Brown, Abolitionist: The Man Who Killed Slavery, Sparked the Civil War, and Seeded Civil Rights* (New York: Vintage Books, 2005) 125–32; and Daegan Miller, "At Home in the Great Northern Wilderness: African Americans and Freedom's Ecology in the Adirondacks, 1846–1859," *Environmental Humanities* 2 (2013): 117–46. Amy Godine curated an exhibition called "Dreaming of Timbuctoo," along with Martha Swan, in 2000, and is currently at work on a manuscript that promises to be the first thoroughly researched exploration of Timbuctoo. See Amy Godine, "Forty Acres and a Vote," *Adirondack Life* 33, no. 6 (Sept.–Oct. 2001): 46–53, and http://adkhistorycenter.org/edu/dot.html. Finally, in 2009, SUNY Potsdam archaeologist Hadley Kruczek-Aaron began excavating the plot of Lyman Epps's home and has begun to write up her findings. See her "Race and Remembering in the Adirondacks: Accounting for Timbuctoo in the Past and the Present," in *The Archaeology of Race in the Northeast*, ed. Christopher N. Matthews and Allison Manfra McGovern (Gainesville: University Press of Florida, 2015), 134–49.

33 Smith owned land in eight New York counties; by far, the majority of land he gave away was in the "Old Military Tract" in Essex and Franklin counties. Smith also paid the cost of drawing up and delivering the deed. Rev. Theodore S. Wright, *An Address to the Three Thousand*, 7, back page; "Distribution of Lands to Colored Men; Begun in 1846," Gerrit Smith Papers, Special Collections Research Center, Syracuse University, vol. 88; and Stauffer, *Black Hearts of Men*, 138.

34 There are few biographies of Gerrit Smith largely because his handwriting was nearly unreadable. I've relied on transcendentalist O. B. Frothingham's edition, despite its overly triumphant tone, as well as Ralph Harlow's, despite its generally sarcastic and negative one. I've also relied heavily on John Stauffer's *Black Hearts of Men*. Frothingham, *Gerrit Smith*, 6–7, 21–22; Harlow, *Gerrit Smith*; and Stauffer, *Black Hearts of Men*.

35 Come-outerism initially referred to those who left established churches. But Lewis Perry has shown how come-outerism took on an anarchistic edge when it turned into a way of viewing the here and now. See Lewis Perry, *Radical Abolitionism: Anarchy and the Government of God in Antislavery Thought* (Ithaca, NY: Cornell University Press, 1973), esp. chap. 4, "Coming Out of Bondage," 92–128.

36 Lewis Perry points out that there were no self-identified anarchists in the abolition movement (the term had only been given a positive spin, by Pierre-Joseph Proudhon, around 1840). Still, even if they didn't describe themselves as anarchists, their basic tenet, that domination of another human is inherently wrong, is one of the foundations of anarchist thought. Harlow, *Gerrit Smith*, 137–47; Frothingham, 181–91; Perry, *Radical Abolitionism*, x, 9, 11, 16–17, and esp. chap. 3, "Nonresistant Anarchism and Anti-

slavery," 55–91, and chap. 6, "The Politics of Anarchy," 158–87. For a brief description of abolitionist politics in New York, see Leslie M. Harris, *In the Shadow of Slavery: African Americans in New York City, 1626–1863* (Chicago: University of Chicago Press, 2003), 202, 210–11, 220–21.

37 Indeed, the late 1830s and early 1840s saw Smith urging more political participation: in 1840 he and other abolitionists formed their own influential but short-lived political party, the Liberty Party. Smith even ran for president in 1848. In 1853, he ran as an Independent for Congress and won, but he resigned in August 1854, tired of congressional deadlock. Perry, *Radical Abolitionism*, 39–46, 184–85; and Stauffer, *Black Hearts of Men*, 4. For a good narrative of the Liberty Party, see Sernett, *North Star Country*, esp. chap. 5, "Bible Politics," 104–28; and for the radical Democrats, the Free Soil Party, and the Liberty Party and Liberty League, see Jonathan H. Earle, *Jacksonian Antislavery & the Politics of Free Soil, 1824–1854* (Chapel Hill: University of North Carolina Press, 2004).

38 Gerrit Smith, "Gerrit Smith's Constitutional Argument," *LT*, Nov. 1, 1844.

39 Thomas Summerhill, *Harvest of Dissent: Agrarianism in Nineteenth-Century New York* (Urbana: University of Illinois Press, 2005), esp. chap. 3, "The Anti-Renters: Agrarians and the Politics of Faction," 60–88; and Stauffer, *Black Hearts of Men*, 136–37.

40 For more on Evans, one of the key radical northern Democrats marrying abolitionism to land reform, see Earle, *Jacksonian Antislavery*, 13, and esp. chap. 1, "Dissident Democrats in the 1830s," and chap. 2, "Set Down Your Feet, Democrats," 17–77.

41 The whole exchange (George H. Evans, "To Gerrit Smith"; Gerrit Smith, "Gerrit Smith's Reply"; and Evans, "Rejoinder to Gerrit Smith") can be found in *People's Rights*, July 24, 1844.

42 There is a large literature touching on agrarianism in the United States. I've relied on Mart A. Stewart, *"What Nature Suffers to Groe": Life, Labor and Landscape on the Georgia Coast, 1680–1920* (Athens: University of Georgia Press, 1996); Jeremy Atack and Fred Bateman, *To Their Own Soil: Agriculture in the Antebellum North* (Ames: Iowa State University Press, 1987); Allan Kulikoff, *The Agrarian Origins of American Capitalism* (Charlottesville: University Press of Virginia, 1992), esp. chap. 2, "The Rise and Demise of the American Yeoman Classes," 34–59, and chap. 3, "The Languages of Class in Rural America," 60–95; Joyce Appleby, *Liberalism and Republicanism in the Historical Imagination* (Cambridge, MA: Harvard University Press, 1992), esp. chap. 10, "The 'Agrarian Myth' in the Early Republic," 253–76; Eric Foner, *Free Soil, Free Labor, Free Men: The Ideology of the Republican Party Before the Civil War* (New York: Oxford University Press, 1995); Sarah T. Phillips, "Antebellum Agricultural Reform, Republican Ideology, and Sectional Tension," *Agricultural History* 71, no. 4 (Autumn 2000): 799–822;

Steven Stoll *Larding the Lean Earth: Soil and Society in Nineteenth-Century America* (New York: Hill and Wang, 2002); and Kimberly K. Smith, *Wendell Berry and the Agrarian Tradition* (Lawrence: University Press of Kansas, 2003), esp. chap. 1, "Agrarian Visions," 11–36. Though it's not quite the same thing as agrarianism, recent work on horticulturalism in the United States has emphasized the connection between cultivating the earth and cultivating oneself. I have been particularly influenced by Philip J. Pauly, *Fruits and Plains: The Horticultural Transformation of America* (Cambridge, MA: Harvard University Press, 2007); Aaron Sachs, *Arcadian America: The Death and Life of an Environmental Tradition* (New Haven, CT: Yale University Press, 2013); Jared Farmer, *Trees in Paradise: A California History* (New York: W. W. Norton & Co., 2013); and Tom Okie, "'Everything Is Peaches Down in Georgia': Culture and Agriculture in the American South" (PhD diss., University of Georgia, 2012).

43 John C. Calhoun, "Speech on the Importance of Domestic Slavery," in *American Political Thought: A Norton Anthology,* ed. Isaac Kramnick and Theodore J. Lowi (New York: W. W. Norton & Co., 2009), 604–5.

44 A rift opened between Northern and Southern agrarians precisely over the question of whether slavery and agrarianism could be compatible. Many Northern agrarians believed that scientific agriculture could work only when land and labor was free. This is where Valenčius's work, though it tends to focus on material health, can be extended to see how healthy landscape could help engender a sort of ethical and social health. See Foner, *Free Soil*, esp. chap. 2, "The Republican Critique of the South," 40–72; Phillips, "Antebellum Agricultural Reform," 808–9, 822; and Valenčius, *Health of the Country*, 3, 229–58, 230, and chap. 8, "Racial Anxiety," 229–58.

45 Cassius M. Clay, "The Voice of the Slaveholder," *LT*, Jan. 1, 1844.

46 See Stoll, *Larding the Lean Earth*, esp. chap. 2, 69–169; Edward Baptist, *The Half Has Never Been Told: Slavery and the Making of American Capitalism* (New York: Basic Books, 2014); Stewart, *"What Nature Suffers to Groe,"* 88, 147–48, passim; Kimberly K. Smith, *African American Environmental Thought: Foundations* (Lawrence: University Press of Kansas, 2007), 8, 18, 154, 157; and Nicolas W. Proctor, *Bathed in Blood: Hunting and Mastery in the Old South* (Charlottesville: University Press of Virginia, 2002) 61, 72. Karl Marx nearly made the same point: capital, he wrote, maximizes profit by "shortening the life of labour-power [i.e., workers], in the same way as a greedy farmer snatches more produce from the soil by robbing it of its fertility." Karl Marx, *Capital* (London: Penguin Books, 1990), 1:376.

47 Christopher Clark notes that between 1800 and 1914 at least 260 communes were formed in the United States, and a great majority of them began in the 1840s. Christopher Clark, *The Communitarian Moment: The Radical Challenge of the Northampton Association* (Ithaca, NY: Cornell University Press,

1995), 10. Donald E. Pitzer, "The New Moral World Order of Robert Owen and New Harmony," in *America's Communal Utopias*, ed. Donald E. Pitzer (Chapel Hill: University of North Carolina Press, 1997), 89–100, 106.

48 The founder of Brook Farm, George Ripley, was a radical abolitionist and a member of Garrison's New England Non-Resistance Society, as was Bronson Alcott of Fruitlands. Sojourner Truth joined the Northampton Association, along with her fellow black abolitionist Samuel Ruggles. Frederick Douglass was a vocal supporter of the experiment, and he also stopped by Hopedale in 1842, where he "moved and melted" the hearts of Hopedale's members. In 1845, Rosetta Hall, "a protégé of Frederick Douglass," and an escaped slave made Hopedale her home for a while. Escaped slaves Ellen and William Craft spent time at Hopedale, as did the infamous slave rescuer, Jonathan Walker, known as "The Man with the Branded Hand." Margaret Washington, *Sojourner Truth's America* (Urbana: University of Illinois Press, 2009), esp. chap. 10, "A Holy City: Sojourner Truth and the Northampton Community," 156–74; Carl J. Guarneri, *The Utopian Alternative: Fourierism in Nineteenth-Century America* (Ithaca, NY: Cornell University Press, 1991), 2–4, 44–59; Ann C. Rose, *Transcendentalism as a Social Movement, 1830–1850* (New Haven, CT: Yale University Press, 1981), 117–61; Philip F. Gura, *American Transcendentalism: A History* (New York: Hill and Wang, 2007), 150–68; Richard Francis, *Fruitlands: The Alcott Family and Their Search for Utopia* (New Haven, CT: Yale University Press, 2010), 1, 281; Clark, *Communitarian Moment*, 12–14, 31, 56–57, 71, 73–74, 99; Adin Ballou, *History of the Hopedale Community: From Its Inception to its Virtual Submergence in the Hopedale Parish* (Lowell, MA: Thompson & Hill—The Vox Populi Press, 1897), 10–11, 77–143; and Perry, *Radical Abolitionism*, 144. For a history of the highly visible African American utopias, see William H. Pease and Jane H. Pease, *Black Utopia: Negro Communal Experiments in America* (Madison: State Historical Society of Wisconsin, 1963).

49 As Henri Lefebvre has argued, spaces have histories and bear upon them the traces of social relations. But spaces also have power and help to shape societies. We might call one such way of reckoning space an abolitionist geography, as does LeMenager; and if there's an abolitionist geography, then there is also a landscape of slavery—and of freedom. Henri Lefebvre, *The Production of Space*, trans. Donald Nicholson-Smith (Malden, MA: Blackwell Publishing, 1991); Stephanie LeMenager, *Manifest and Other Destinies: Territorial Fictions of the Nineteenth-Century United States* (Lincoln: University of Nebraska Press, 2004), 177–88, and Stephanie LeMenager, "Marginal Landscape: Revolutionary Abolitionists and Environmental Imagination," *Interdisciplinary Literary Studies* 7, no. 1 (Fall 2005): 49–56.

50 John L. Thomas, "Antislavery and Utopia," in *The Antislavery Vanguard: New Essays on the Abolitionists*, ed. Martin Duberman (Princeton, NJ: Princeton University Press, 1965), 258, 254–59. Monique Allewaert's recent work

suggests that "in many cases is was subaltern persons and those in close proximity to them who forged the first responses to the conjoined problems of economic and environmental exploitation." Though she privileges the tropics as the superlative environment for social and ecological critiques, I think Allewaert's work can very profitably help inform what was happening in the Adirondacks. See Monique Allewaert, *Ariel's Ecology: Plantations, Personhood, and Colonialism in the American Tropics* (Minneapolis: University of Minnesota Press, 2013), 18.

51 In fact, it was Noyes who initially turned Garrison down the path to anarchistic thinking, partially by means of a long letter in which he compared the US government to "a fat libertine flogging Negroes and torturing Indians." Perry, *Radical Abolitionism*, 63–69; and Stauffer, *Black Hearts of Men*, 131–33.

52 Ballou, *History of the Hopedale Community*, 47.

53 Besides the agricultural Adirondack community, Smith set up an urban abolitionist commune on part of his inheritance in western New York, near the town of Florence. The Florence community was apparently part of a larger scheme, open not only to New Yorkers but also to African Americans throughout the Northeast. Finally, by 1849 and 1850, Smith was also giving land away to more than one thousand needy white residents of New York. Most of the land was not in the Adirondacks, however, and was of lesser quality. "Settlements for Colored Men," *NS*, Dec. 8, 1848; "A New Settlement," *NS*, Dec. 22, 1848; "Florence Settlement," *NS*, Feb. 23, 1849; "New Bedford, March 22, 1848," *NS*, Mar. 30, 1849; see also box 98, index to ledger A, Florence lands undated folder, and box 106, Lands in the town of Florence, Oneida County folder, GSP.

54 "Extract," *NS*, Apr. 21, 1848.

55 Richard White, "'Are You an Environmentalist or Do You Work for a Living?': Work and Nature," in *Uncommon Ground: Rethinking the Human Place in Nature*, ed. William Cronon (New York: W. W. Norton & Co., 1996), 171.

56 The sources are unclear on the point of Smith's attitude toward female grantees, but it does seem that Smith was initially open to deeding African American women land. He later reconsidered, and instead gave eligible women $50 each. See John Cochrane, Daniel C. Eaton, and George H. Evans, List of Beneficiaries, n.d. [ca. 1850], box 145, Gifts of Land and Money to Negroes folder, GSP; Asa B. Smith to Gerrit Smith, n.d. [ca. 1850], box 145, Gifts of Land and Money to Negroes folder, GSP; John Cochrane, Isaac Hopper, D. C. Easton, and William Kinney to Gerrit Smith, Jan. 2, 1850, box 145, Gifts of Land and Money to Negroes folder, GSP.

57 Evans would become one of the second-tier agents who helped draw up names of potential grantees. There are a number of letters in the Gerrit Smith Papers at Syracuse University's Special Collections Research Letter in which Evans is one of the signatories (some of which I've cited here).

58 Henry Highland Garnet to Gerrit Smith, Sept. 22, 1848, box 20, Garnet, Henry Highland 1845–73 folder, GSP; Charles B. Ray and James McCune Smith to Gerrit Smith, Jan. 20, 1848, box 31, Ray, Charles Bennett Incoming Corres., 1847–1873, GSP.
59 Sernett argues that though "movement abolitionism," those attuned to Garrisonian critiques, has traditionally been told as a tale of white activism, African Americans played a large role, and that, especially in upstate New York, their voices contributed to a richly potent brand of antislavery resistance. Sernett, *North Star Country*, xix–xx, 301–4; Wright et al., *An Address to the Three Thousand Colored Citizens*, 9.
60 Wright et al., *An Address to the Three Thousand Colored Citizens*, 10.
61 James McCune Smith to Gerrit Smith, Dec. 17, 1846, box 34, Smith, James McCune Incoming Corres. folder, GSP.
62 Wright et al., *An Address to the Three Thousand Colored Citizens*, 10.
63 For work that explores an African American environmental history and ecocriticism, and the connections of environment and race, see, besides works cited previously, Melvin Dixon, *Ride Out the Wilderness: Geography and Identity in Afro-American Literature* (Urbana: University of Illinois Press, 1987); Judith A. Carney, *Black Rice: The African Origins of Rice Cultivation in the Americas* (Cambridge, MA: Harvard University Press, 2001); Alison H. Deming and Lauret E. Savoy, eds., *The Colors of Nature: Culture, Identity, and the Natural World* (Minneapolis: Milkweed Editions, 2002); Elizabeth D. Blum, "Power, Danger, and Control: Slave Women's Perceptions of Wilderness in the Nineteenth Century," *Women's Studies* 31 (2002): 247–65; Carolyn Merchant, "Shades of Darkness: Race and Environmental History," *Environmental History* 8 (July 2003): 380–94; Dianne D. Glave, "A Garden So Brilliant with Colors, So Original in Its Design: Rural African American Women, Gardening, Progressive Reform, and the Foundation of an African American Environmental Perspective," *Environmental History* 8 (July 2003): 395–411; Dianne D. Glave and Mark Stoll, eds., *"To Love the Wind and Rain": African Americans and Environmental History* (Pittsburgh, PA: University of Pittsburgh Press, 2006); Dianne D. Glave, *Rooted in the Earth: Reclaiming the African American Environmental Heritage* (Chicago: Lawrence Hill Books, 2010); Smith, *African American Environmental Thought*; Camille T. Dungy, ed., *Black Nature: Four Centuries of African American Nature Poetry* (Athens: University of Georgia Press, 2009); Kimberly N. Ruffin, *Black on Earth: African American Ecoliterary Traditions* (Athens: University of Georgia Press, 2010); and Carolyn Finney, *Black Faces, White Spaces: Reimagining the Relationship of African Americans to the Great Outdoors* (Chapel Hill: University of North Carolina Press, 2014).
64 There's an enormous range of scholarship that one could cite here. I've leaned heavily on Stephen Jay Gould, *The Mismeasure of Man*, rev. ed. (New York: W. W. Norton & Co., 1996); Ronald Takaki, *Iron Cages: Race and Cul-*

ture in 19th-Century America, rev. ed. (New York: Oxford University Press, 2000); Bruce Dain, *A Hideous Monster of the Mind: American Race Theory in the Early Republic* (Cambridge, MA: Harvard University Press, 2002); Anne Fabian, *The Skull Collectors: Race, Science, and America's Unburied Dead* (Chicago: University of Chicago Press, 2010); Karen E. Fields and Barbara J. Fields, *Racecraft: The Soul of Inequality in American Life* (New York: Verso, 2012); and Baptist, *The Half Has Never Been Told*.

65 Karen Fields and Barbara Fields write that "racecraft" is "the process of reasoning" that makes race plausible, even in the face of overwhelming counter-evidence, a process that seems to turn subjective social suspicion into objective natural fact. See Fields and Fields, *Racecraft*, 5–11, 16–24.

66 Thomas Jefferson, *Notes on the State of Virginia* in *Thomas Jefferson: Writings* (New York: Library of America, 1984), 264.

67 Ibid., 264–66.

68 Ibid., 270. Bruce Dain argues that Jefferson's theories of racial stability were out of step with the near-evolutionary thinking of the nineteenth century, and so Dain disagrees with the common argument that Jefferson is the starting point for American scientific racism. But one's reasoning need not be fashionable for one's thinking to prove influential. Jefferson's racial *science* might not have caught on, but his racecraft did. Dain, *Hideous Monster of the Mind*, esp. chap. 1, "The Face of Nature," 1–39.

69 Samuel George Morton, *Crania Americana; or, A Comparative View of the Skulls of Various Aboriginal Nations of North and South America, to which is Prefixed an Essay on the Varieties of the Human Species* (Philadelphia: J. Dobson, 1839), 5–7.

70 Fabian, *Skull Collectors*, 82–83.

71 For Louis Agassiz's views on natural history, race, and environment, see his "Geographical Distribution of Animals," *Christian Examiner and Religious Miscellany* 48, no. 2 (Mar. 1850), and "The Diversity of Origin of the Human Races," *Christian Examiner and Religious Miscellany* 49, no. 1 (July 1850); Louis Menand, *The Metaphysical Club: A Story of Ideas in America* (New York: Farrar, Straus & Giroux, 2001), 102–12; and Fabian, *Skull Collectors*, 112–15.

72 Stauffer argues that Gerrit Smith and James McCune Smith—as well as John Brown and Frederick Douglass—were remarkable for being among the very few Americans who could blur the color line. One of the ways they did so was to emphasize something common to all humans: labor, and in this case, the most idealized form of labor in the United States, husbandry. Valenčius points out that to cultivate, the act of the husbandman, was often explicitly an activity of healing, of bringing something to its fullest potential. And so there's a clear link between the work of cultivating and the work of healing a nation eroded by racial discord. Stauffer, *Black Hearts of Men*, 2, 14, 19, 38, passim; and Valenčius, *Health of the Country*, 192.

73 "Correspondence," *NS*, Jan. 7, 1848.
74 Old Military Tract, box 98, Description of Lots and Acreage Books folder, vol. 47d, GSP.
75 Godine, "Forty Acres and a Vote," 52; "Convention of Gerrit Smith's Grantees," *NE*, Feb. 11, 1847; "To Hon. Gerrit Smith," *NE*, Apr. 15, 1847; "Movements of the Grantees of the City of Troy," *NS*, Nov. 10, 1848.
76 "Gerrit Smith's Lands," *NS*, Mar. 24, 1848.
77 Willis Hodges, *Free Man of Color: The Autobiography of Willis Augustus Hodges*, ed. William B. Gatewood (Knoxville: University of Tennessee Press, 1982), xliv, 77–80.
78 Clark, *Communitarian Moment*, 58; "Correspondence," *NS*, Jan. 7, 1848; "The Smith Lands," *NS*, Feb. 18, 1848.
79 For a detailed exploration of the Adirondack tanning industry, see Barbara McMartin, *Hides, Hemlocks and Adirondack History* (Utica, NY: North Country Books, 1992), 48, 50, 106.
80 For statistics on natural resource use in the Adirondacks, see Barbara McMartin, *The Great Forest of the Adirondacks* (Utica, NY: North Country Books, 1994), esp. chap. 5, "The Adirondacks from 1850–1890: Forests as Industry," 29–74.
81 The term *lumber frontier* comes from Thomas R. Cox, who writes that the first lumber frontier in the US Mid-Atlantic region was born at Glens Falls, New York—at the southern end of the Adirondacks—in 1763. Cox, *Lumberman's Frontier*, 94–100; Williams, *Americans & Their Forests*, 96; Schneider, *Adirondacks*, 210; Cox, *This Well-Wooded Land*, 117; William F. Porter, "Forestry in the Adirondacks," in *The Great Experiment in Conservation: Voices from the Adirondack Park*, ed. William F. Porter, Jon D. Erickson, and Ross S. Whaley (Syracuse, NY: Syracuse University Press, 2009), 103–8; Schneider, *Adirondacks*, 202, 216.
82 Jones writes that, "for a people who have known the agony of a system in which family members could be routinely sold away from one another—wives from husbands, children from mothers—land stood as one of the only tangible possessions that could not be easily confiscated. For African Americans, the attainment of land was a priceless step toward self-sufficiency and security." This is exactly where African American history and the intellectual history of agrarianism and horticulturalism intersect. Jones, "They Called It Timbuctoo," 32.
83 "New York, May 18, 1848," *NS*, Jan. 12, 1849.
84 Quoted in Don Papson, "The John Thomas Story: From Slavery in Maryland to American Citizenship in the Adirondacks," in *Lake Champlain Weekly*, Oct. 18, 2006. I owe a debt of thanks to Don Papson for sending me his articles.
85 "Meeting of the Rochester Grantees," *NS*, Dec. 15, 1848.
86 "Convention of Colored People," *NS*, Oct. 20, 1848.

87 This has been one of the central, axiomatic principles of the work on wilderness stretching all the way back to Frederick Jackson Turner's famed frontier thesis. For more modern elaborations, see, among many others, Miller, "At Home in the Great Northern Wilderness"; Finney, *Black Spaces, White Faces*, esp. the introduction, 1–20; Kevin DeLuca and Anne Demo, "Imagining Nature and Erasing Class and Race: Carleton Watkins, John Muir, and the Construction of Wilderness," in *The Wilderness Debate Rages On: Continuing the Great New Wilderness Debate*, ed. J. Baird Callicott and Michael P. Nelson (Athens: University of Georgia Press, 2008); Glave, *Rooted in the Earth*; Deming and Savoy, *Colors of Nature*; and Smith, *African American Environmental Thought*.

88 One might also question how homogeneous the wilderness ideas championed by the traditional patron saints of American wilderness were. Even lumping together Thoreau and Emerson is to group dissimilar thinking under the same banner.

89 Cronon, "Trouble with Wilderness," 85.

90 "I'm Going Home," in *Slave Songs of the United States* (1867; repr., New York: Peter Smith, 1951), 84. Dialect in the original.

91 Dixon, *Ride Out the Wilderness*, 14, 17, 20, 27. Besides the wilderness, Dixon mentions that the mountaintop and the valley crop up the most frequently in African America green culture — of course, the Adirondacks are characterized by their steep mountains and deep valleys. For a rethinking of the role of wilderness in African American intellectual history, see Kimberly K. Smith, "What Is Africa to Me? Wilderness in Black Thought, 1860–1930," in *The Wilderness Debate Rages On: Continuing the Great New Wilderness Debate*, ed. J. Baird Callicott and Michael P. Nelson (Athens: University of Georgia Press, 2008), 300–324.

92 Kimberly N. Ruffin has argued that African Americans historically "forged identities as ecological participants based on their work rather than a privileged position in the social fabric." Looking at black intellectual history, then, is one way to answer Richard White's persuasive challenge to environmental historians to start recognizing work as one of our most important daily interactions with the natural world. Furthermore, Smith's identification of a tradition that she calls black agrarianism is particularly useful here, especially her argument that black agrarians "fused the abolitionists' north-south moral geography with the sacred landscape of the slave spirituals . . . [creating] a moral landscape with both political and spiritual meaning." See Ruffin, *Black on Earth*, 28, 29, 40, 42, 54; White, "'Are You an Environmentalist or Do You Work for a Living?'" 171–72, 173; and Smith, *African American Environmental Thought*, 56, 58.

93 It's just such a definition that Kimberly K. Smith ascribes to what she calls the "the black concept" and the "black wilderness tradition." Leave aside for the moment that Smith's definition equally well describes the utopian

socialist communes of the 1840s and 1850s and even sounds like a good brief of Thoreau's cartography; what Smith identifies as key to a black wilderness is the relationship of a community to the land, one that theorizes a living wilderness capable of staking a claim in a people's memory. Smith, "What Is Africa to Me?" 301.

94 Stauffer, *Black Hearts of Men*, 5–6, 15, 66.
95 Dain, *A Hideous Monster of the Mind*, esp. chap. 8, "Effacing the Individual," 227–63.
96 James McCune Smith to Gerrit Smith, Dec. 28, 1846, box 34, Smith, James McCune Incoming Corres. folder, GSP.
97 Ibid.
98 Ibid.
99 Ibid.
100 Names illegible [C. S. Morton, Benjamin Latimore, P. W. Grommell, and J. P. Anthony, Richard Thompson] to Gerrit Smith, Nov. 4, 1846, box 145, Gifts of Land and Money to Negroes folder, GSP.
101 Charles B. Ray and James McCune Smith to Gerrit Smith, July 27, 1847, box 31, Ray, Charles Bennett Incoming Corres. 1847–1873 folder, GSP.
102 Charles B. Ray to Gerrit Smith, May 24, 1847, box 31, Ray, Charles Bennett Incoming Corres. 1847–1873 folder, GSP; Richard Henry Dana, *The Journal of Richard Henry Dana Jr.*, ed. Robert F. Lucid (Cambridge, MA: Belknap Press of Harvard University Press, 1968), 1:373; and Charles B. Ray and James McCune Smith to Gerrit Smith, July 27, 1847.
103 James McCune Smith to Gerrit Smith, Mar. 27, 1848, box 34, Smith, James McCune Incoming Corres. folder, GSP; James McCune Smith to Gerrit Smith, May 12, 1848, box 34, Smith, James McCune Incoming Corres. folder, GSP.
104 James McCune Smith to Gerrit Smith, July 7, 1848, box 34, Smith, James McCune Incoming Corres. folder, GSP.
105 James McCune Smith to Gerrit Smith, Feb. 6, 1850, box 34, Smith, James McCune Incoming Corres. folder, GSP.
106 Ibid.
107 John Thomas to Gerrit Smith, Bloomingdale, Essex County, NY, Aug. 26, 1872, microfilm, reel 18 (Glen Rock, NJ: Microfilming Corp. of America, 1975), GSP.
108 McCune Smith to Gerrit Smith, Feb. 6, 1850.
109 Franklin Sanborn's conclusion—that "there was no opening in the woods of Essex for waiters, barbers, coachmen, washer-women, or the other occupations for which negroes had been trained"—a conclusion that was advanced to bolster Sanborn's hagiographic take on Brown, has become the dominant one. See F. B. Sanborn, ed., *The Life and Letters of John Brown, Liberator of Kansas, and Martyr of Virginia* (New York: Negro Universities Press, 1969), 97.

110 "Mr. Waite J. Lewis and the Smith's Lands," *NS*, Feb. 16, 1849.
111 The longest lived was Lyman Epps Jr., who was born in the Adirondacks and whose father, Lyman Epps, was one of the original grantees and helped build the Brown house (he wrote his name on a board that can still be seen in the attic of the Brown farm). Epps Jr. died in 1942. And there are still descendants of the pioneers who live in the Adirondack region. See Don Papson's series of articles, "The John Thomas Story: From Slavery in Maryland to American Citizenship in the Adirondacks," *Lake Champlain Weekly*, Oct. 18, 2006; Oct. 25, 2006; Nov. 1, 2006; and especially Nov. 8, 2006; and Schneider, *Adirondacks*, 186n1.
112 John Brown apparently kept a little ledger book for each of his children: one column was for their minor daily transgressions; the other was for how many lashes Brown would lay on their backs as punishment:

John Jr.,
    For disobeying mother . . . . 8 lashes
    " unfaithfulness at work . . . . 3 "
    " telling a lie . . . . 8 "

Every once in a while, Brown came around with a switch in one hand, and his little ledger in the other, ready to settle up. Oddly enough, Brown's system was the exact same one used by Southerners—they called it the pushing system—to drive their slaves to pick ever more cotton. See Sanborn, *Life and Letters of John Brown*, 91–93; Reynolds, *John Brown*, 41, and Baptist, *The Half Has Never Been Told*, esp. chap. 4, "Left Hand, 1805–1861," 111–44.
113 Quoted in Sanborn, *Life and Letters of John Brown*, 97. Smith had probably been on Brown's radar for nearly a decade. Smith had given the antislavery Oberlin College a large tract of land in Virginia, and in 1840, Brown, broke even then, angled to survey it. Reynolds, *John Brown*, 74–75; and Stauffer, *Black Hearts of Men*, 120.
114 Quoted in Harlow, *Gerrit Smith*, 246.
115 Ruth Thompson, quoted in Sanborn, *Life and Letters of John Brown*, 99.
116 In January 1849, Brown wrote to Hodges that "Mr. Pennington is about to start for the West Indies and for England, to collect means to aid in building up the infant colony on the Smith lands." "John Brown in Essex County," *Evening Post* (New York), Dec. 20, 1859.
117 It's difficult to pinpoint exactly when Brown was in the Adirondacks. I've cobbled my chronology together from Sanborn, *Life and Letters of John Brown*, 97–115; Donaldson, *History of the Adirondacks*, 2:3–12; Stauffer, *Black Hearts of Men*, 168–74; and Reynolds, *John Brown, Abolitionist*, 89, 94, 125–37, 233.
118 John Brown to John Brown Jr., June 1854, quoted in Sanborn, *Life and Letters of John Brown*, 105.

119 The literature on Brown is immense, so I have chosen the most influential examples of how Brown scholars have written about the Adirondacks. Reynolds writes, "[Brown's] community, like his other projects, was destined to fail economically," but the settlements were never his idea in the first place, and there's no reason to make Brown the major protagonist in this history. One of Brown's earliest biographers, Oswald Garrison Villard, constructs his narrative to reflect Brown's selflessness in moving to the Adirondacks, and there's really no sense that the pioneers had any agency themselves. And Stephen B. Oates, although he does a remarkable job in capturing Brown's desperation in the late 1840s, also chalks the entire Adirondack experiment up to Brown: "As their new leader he hoped to teach these 'poor despised Africans' how to farm and better themselves, and he would also tech them to fear God. If it were the will of Providence, he would develop North Elba from the disorganized place it was now into a model Negro community that would stand as an example to the world." See Reynolds, *John Brown, Abolitionist*, 126; Oswald Garrison Villard, *John Brown: A Biography Fifty Years After* (Gloucester, MA: Peter Smith, 1965), 71–76; Stephen B. Oates, *To Purge This Land with Blood: A Biography of John Brown*, 2nd ed. (Amherst: University of Massachusetts Press, 1984), 66.
120 James McCune Smith to Gerrit Smith, Feb. 6, 1850.
121 Stauffer, *Black Hearts of Men*, 172, 184.
122 Too infrequently have historians recognized the great intellectual costs of Brown's victory, and too frequently have historians held Brown up as a shining example of revolutionary commitment, the symbol of someone willing to live by his convictions. The subtitle of Reynolds's book is *The Man Who Killed Slavery, Sparked the Civil War, and Seeded Civil Rights*—who doesn't want to stand in that tradition? I think that the point of deifying Brown is to let ourselves off the hook: he acted so we don't have to. Even so, I'm not sure there's anything particularly worthwhile in Brown's example. Great tragedy has been brought upon the world by those who were willing to enforce God's law, as revealed to them only. While Brown's end—destroying slavery—was a social good, I can't find a way to accept his method, and I am one of those who do not think that the ends justify the means. This is why I think that it's a double tragedy that the Adirondack experiment has been forgotten, for its simplicity is the root of its radical proposition: as a historical example and challenge, it demands our best action rather than lazily relying on bludgeoning our opponents into acquiescence. It's powerful in the way that all truly radical movements are: it asks of us so little, only that we work together. It was a demand that asked too much of Brown, and he buried it, too. My thinking on the history of Brown owes much to Michel-Rolph Trouillot's *Silencing the Past: Power and the Production of History* (Boston: Beacon Press, 1995).
123 Lewis Perry notes that nonresistance was one of the casualties of Harper's

Ferry, and that after 1859 "almost no nonresistant voice remained to be raised against force and violence. . . . Once [radical abolitionists] had criticized men and institutions for relying on the implicit threat of violence; now their criticism was directed against those who did not live up to their standards of violence that nonresistants imputed to them." Perry, *Radical Abolitionism*, 259, 260–61. For similar analyses, see also Thomas, "Antislavery and Utopia," 265, 268; Gura, *American Transcendentalism*, 304–6; Rose, *Transcendentalism as a Social Movement*, 207–8, 218–25; Guarneri, *Utopian Alternative*, esp. chap. 15, "Fourierism and the Coming of the Civil War," 368–83.

124 Terrie's two books, *Forever Wild* and *Contested Terrain*, Graham's *Adirondack Park*, Jacoby's *Crimes against Nature*, and Schneider's *Adirondacks* do a very good job, when taken together, of charting the major developments in the Adirondack wilderness-as-empty story.

125 William H. H. Murray, *Adventures in the Wilderness; or, Camp-Life in the Adirondacks* (Boston: Fields, Osgood, & Co., 1869), 43.

126 W. J. Stillman, *The Autobiography of a Journalist* (London: Grant Richards, 1901), 1:209.

127 Ibid., 1:167.

128 Ibid., 2:211.

129 My thanks to Phil Terrie for pointing out Raymond's journalistic connections.

130 J. T. Headley, *The Adirondack; or, Life in the Woods* (New York: Baker and Scribner, 1849), dedication.

131 Ibid., i.

132 Ibid., iii.

133 Ibid., 115–16.

134 Ibid., 168.

135 Stillman, *Autobiography of a Journalist*, 1:179–80.

136 Headley, *Adirondack*, 198; S. H. Hammond, *Wild Northern Scenes; or Sporting Adventures with the Rifle and the Rod* (New York: Derby & Jackson, 1857), 29.

137 Hammond, *Wild Northern Scenes*, 310–13.

138 Ibid., 314–16.

139 Ibid., 316, 341.

140 Schneider, *Adirondacks*, esp. chap. 15, "At Play in the Great Longhouse," 175–91; chap. 21, "Birth of a Great Camp," 241–58; and chap. 22, "Haute Rustic," 259–76; and Terrie, *Contested Terrain*, 61–74.

141 Schneider, *Adirondacks*, 168; Murray, *Adventures in the Wilderness*, 12–14.

142 Robert Louis Stevenson was one of Trudeau's patients, as was Alfred L. Donaldson, the historian whom I've relied on throughout this chapter. Schneider, *Adirondacks* 167–168; and Terrie, *Contested Terrain*, 65. Susan Sontag writes brilliantly of the culture that grew up around tuberculosis, generated, especially, by the seeming randomness and incurability of the

disease. She sees consumption as the first archetypical disease of modernity, and "with TB . . . the idea of individual illness was articulated . . . in the images that collected around the disease one can see emerging a modern idea of individuality." Susan Sontag, *Illness as Metaphor and AIDS and Its Metaphors* (New York: Picador, 1989), 30.

143 Gregg Mitman has recently explored the intersection of environmental and medical history by focusing on allergies and air quality, and how their perception led to the creation of particular landscapes. See Gregg Mitman, *Breathing Space: How Allergies Shape Our Lives and Landscapes* (New Haven, CT: Yale University Press, 2007), esp. chap. 1, "Hay Fever Holiday," 10–51.

144 Marc Cook, *The Wilderness Cure* (New York: William Wood & Company, 1881), 17.

145 Ibid., 83–84.

146 Cronon, "Trouble with Wilderness," 81.

147 In a similar vein, Paul Sutter has written of revisionist wilderness criticism: "Rather than attending to the complex history of wilderness advocacy, wilderness critics have conveniently or inadvertently lumped wilderness advocates together, intimating that all hold to an idea of wilderness that is by turns ecologically naïve, dispossessive, class-biased, consumerist, and hopelessly separated from concerns for social justice. These critics have abstracted the wilderness idea from politics, reified it, and built by logic and selectivity a profile of advocacy that misses complexity, contingency, and context." See Paul Sutter, *Driven Wild: How the Fight against Automobiles Launched the Modern Wilderness Movement* (Seattle: University of Washington Press, 2002), 13. For an alternate definition of wilderness, see Rebecca Solnit, *A Field Guide to Getting Lost* (New York: Penguin Books, 2005), 89; and for a very different history, see John Brinckerhoff Jackson, "Beyond Wilderness," in *A Sense of Place, a Sense of Time* (New Haven, CT: Yale University Press, 1994), 71–92.

148 McCune Smith to Gerrit Smith, Feb. 6, 1850; "The Smith Land," *NS*, Feb. 18, 1848; Headley, *Adirondack*, 198; Hammond, *Wild Northern Scenes*, 84; John A. Hows, *Forest Pictures in the Adirondacks: With Original Poems by Alfred B. Street* (New York: James G. Gregory, 1864), 1; Murray, *Adventures in the Wilderness*, 195.

149 Homer D. L. Sweet, *Twilight Hours in the Adirondacks: The Daily Doings and Several Sayings of Seven Sober, Social, Scientific Students in the Great Wilderness of Northern New York* (Syracuse, NY: Wynkoops & Leonard, 1870), ll. 19–20; Cook, *Wilderness Cure*, 43; H. Perry Smith, *The Modern Babes in the Woods; or, Summerings in the Wilderness. To Which is Appended a Reliable and Descriptive Guide to the Adirondacks by E. R. Wallace* (Hartford, CT: Columbian Book Company, 1872), 24.

150 As Conevery Bolton Valenčius has argued, "'Settling' meant, in all practical terms, cultivating." One can even trace this back to the Puritans. In 1672,

Thomas Shepard, an important Puritan minister, delivered the annual Election Day sermon, and in it he defined wilderness in five ways: "a desolate, solitary place, without Inhabitants": "a place uncultivated," a place where "there is not a beaten path," a place where "there is not only want of comfort, but there is danger as to many positive evils," and "*A wilderness is not hedged in, nor fenced about*" (italics in original). Only one of Shepard's definitions, then, turns on emptiness, and it seems clear that inhabitation, for Shepard and his congregants, meant cultivation. See Valenčius, *Health of the Country*, 193; and Shepard, quoted in Steven D. Neuwirth, "The Images of Place: Puritans, Indians, and the Religious Significance of the New England Frontier," *American Art Journal* 18, no. 2 (Spring 1986): 43. See also Jared Farmer, *On Zion's Mount: Mormons, Indians, and the American Landscape* (Cambridge, MA: Harvard University Press, 2008), 50.

151 Schneider, *Adirondacks*, 122–23. For a comprehensive history of Adirondack mining, see Patrick F. Farrell, *Through the Light Hole: A Saga of Adirondack Mines and Men* (Utica, NY: North Country Books, 1996).

152 S. R. Stoddard, *The Adirondacks Illustrated*, 10th ed. (Albany, NY: Charles Van Benthuysen & Sons, 1881), 2, 4, passim.

153 Philosopher Wayne Ouderkirk, among others, has recently taken William Cronon and J. Baird Callicott to task for just such simplification. Ouderkirk argues instead that "we continue to respect the wild because it is simultaneously something of which we are a part and which lives on in its own way." Humans are dialectically intertwined with what we are not, and to collapse the difference, to argue that there is no difference between the human and nonhuman is every bit as homogenizing as arguing that the human and the wild are mutually exclusive categories. See Wayne Ouderkirk, "On Wilderness and People: A View from Mt. Marcy," in *The Wilderness Debate Rages On: Continuing the Great New Wilderness Debate*, ed. J. Baird Callicott and Michael P. Nelson (Athens: University of Georgia Press, 2008), 450, 451, 454.

154 Headley, *Adirondack*, 118. The landscape architect and historian John Dixon Hunt has argued that the Western tradition of landscape architecture inherited from the Renaissance prizes continuity rather than discrete separation, a flowing from first nature (the nonhuman world) through second nature (the human-created landscape) and into third nature (the interventions, like landscaped gardens, that go beyond strictly utilitarian need). See John Dixon Hunt, *In Greater Perfections: The Practice of Garden Theory* (London: Thames & Hudson, 2000), esp. chap. 3, "The Idea of a Garden and the Three Natures," 32–75. See also Massey, *For Space*, 65.

155 I'm riffing here on Val Plumwood's "Wilderness Skepticism and Wilderness Dualism," in which she argues against the received wilderness idea of wilderness as feminine and empty, but *for* the idea of wilderness as feminine and full of difference. Too much of the recent criticism of wilderness, she argues,

either simply flips the dualistic categories, or collapses them altogether. "To overcome this dualism," she writes, "we need to reclaim the ground of continuity, to recognize both the culture which has been denied in the sphere conceived as pure nature, and to recognize the nature which has been denied in the sphere conceived as pure culture. The traditionally dualistic wilderness concept delegitimates both, denying the legitimacy or possibility of the hybrids and boundary crossings which break up the neatly regimented polarity of nature and culture, and which enable wilderness reserves to be understood as part of a continuum." Val Plumwood, "Wilderness Skepticism and Wilderness Dualism," in *The Great New Wilderness Debate*, ed. J. Baird Callicott and Michael P. Nelson (Athens: University of Georgia Press, 1998), 655, 659, 669, 670, 679; and see, more generally, Plumwood, *Feminism and the Mastery of Nature*. I also think that wilderness as a place of connection and of completion is just how Bill McKibben can call the Adirondacks a part of "America's Most Hopeful Landscape." Bill McKibben, *Wandering Home: A Long Walk across America's Most Hopeful Landscape; Vermont's Champlain Valley and New York's Adirondacks* (New York: Crown Publishers, 2005).

156 Indeed, Nicolai Cikovsky has argued that one of the most popular subjects in American landscape art of the mid-nineteenth-century was the home in the wilderness. See Nicolai Cikovsky Jr., "'The Ravages of the Axe': The Meaning of the Tree Stump in Nineteenth-Century American Art," *Art Bulletin* 61, no. 4 (Dec. 1979): 622. For a reading of Cole's *Home in the Woods* that is parallel to mine, see John Conron, *American Picturesque* (University Park: Pennsylvania State University, 2000), 137. And Jerome Tharaud has recently argued that Cole's work operated in and helped develop a cultural "evangelical space" which lends texture, complexity, and ambiguity to the traditional historical argument that landscape painting was simply the artistic lackey of Manifest Destiny. Jerome Tharaud, "Evangelical Space: *The Oxbow*, Religious Print, and the Moral Landscape in America," *American Art* 28, no. 3 (Fall 2014): 52–75. Headley, *Adirondack*, 21, 22–23.

157 According to the 2010 US Census, the median household income for Franklin County was $41,062 and $42,053 for Essex. For comparison, the richest county in the state, Nassau County, had a median household income of $90,294, more than twice the income of both Franklin and Essex counties. See "Small Area Income and Poverty Estimates," *US Census Bureau*, http://www.census.gov/did/www/saipe/county.html (accessed Apr. 22, 2012).

158 Of the state land, 96.4 percent is classified as either wilderness (45.5 percent) or wild forest (50.9 percent). See the Adirondack Park Agency's May 2014 Park Land Use Classification Statistics, available at http://apa.ny.gov/gis/stats/colc201405.htm.

159 The clause covers not only the Adirondacks but also the state-owned lands in the Catskill Park. Graham's *Adirondack Park* is still the indispensible reference for the legal history of the Park. See chap. 15, "Forever Wild," 126–32.

160 Terrie, *Forever Wild*, 100–101.
161 Stradling, *The Nature of New York*, especially his introduction, "Nature Is on the Side of New York," 1–13.
162 Williams, *Americans & Their Forests*, 406; Cox, *This Well-Wooded Land*, 90–93; and especially Graham, *Adirondack Park*, chap. 12, "The City Intervenes," 96–106. Terrie writes that "the aesthetic argument alone could never have led to the preservation of the Adirondacks. Rather, it became just one more weapon in the arsenal of the utilitarian protectors of the watershed." Terrie, *Forever Wild*, 101; for a more general work on the nineteenth-century understanding of the relationship between forest and urban health, see Hou, *City Natural*.
163 Ralph Ellison, *Shadow and Act* (New York: Random House, 1964), 74.
164 For the etymological roots shared by both *tree* and *truth* (via *true*), see *Oxford English Dictionary* online, s.v. "tree" and "true," http://www.oed.com, accessed Apr. 7, 2015.
165 "Gerrit Smith's Lands," *NS*, Mar. 24, 1848.

INTERMISSION

1 Marcel Proust, *Within a Budding Grove* (1919), vol. 1 of *Remembrance of Things Past*, trans. C. K Scott Moncrieff and Terrence Kilmartin (New York: Vintage Books, 1982), 773.
2 Edward William Lane, trans., *The Arabian Nights' Entertainments; or, The Thousand and One Nights* (New York: Tudor Publishing, 1944), 678.
3 Murray, *Adventures in the Wilderness*, 208–11, 214, 215, 216–18, 227–28, 228–29, 230, 231–32.

ACT THREE

1 Thomas Pynchon, *Against the Day* (New York: Penguin Books, 2007), 64.
2 Andrew Joseph Russell, "On the Mountains with the Tripod and Camera," *APB* 1 (1870): 33; Gaston Tissandier, *A History and Handbook of Photography*, ed. J. Thomson (1878; repr., New York: Arno Press, 1973), 154; *Andrew J. Russell: Visual Historian*, U-matic (Brigham Young University Production, 1983). For comprehensive works on Russell, it is best to begin with Susan Danly Walther's dissertation, "The Landscape Photographs of Alexander Gardner and Andrew Joseph Russell" (PhD diss., Brown University, 1983), and her later articles: Susan Danly, "Andrew Joseph Russell's *The Great West Illustrated*," in *The Railroad in American Art: Representations of Technological Change*, ed. Susan Danly and Leo Marx (Cambridge, MA: MIT Press, 1988) and "Photography, Railroads, and Natural Resources in the Arid West:

Photographs by Alexander Gardner and A. J. Russell," in *Perpetual Mirage: Photographic Narratives of the Desert West*, ed. May Castleberry (New York: Whitney Museum of American Art, 1996). Susan E. Williams is one of Russell's earliest and most prolific biographers. See Susan E. Williams, "*The Great West Illustrated*: A Journey across the Continent with Andrew J. Russell," *Streamliner: The Official Publication of the Union Pacific Historical Society* 10, no. 3 (1995): 3–19; "'Richmond Again Taken': Reappraising the Brady Legend through Photographs by Andrew J. Russell," *Virginia Magazine of History and Biography* 110, no. 4 (2002): 437–60; and "The Truth Be Told: The Union Pacific Railroad Photographs of A. J. Russell," *View Camera* (Jan.–Feb. 1996): 36–43. Martha Sandweiss turns to Russell in chapter 5, "'Westward the Course of Empire,' Photography and the Invention of an American Future," of *Print the Legend: Photography and the American West* (New Haven, CT: Yale University Press, 2002), 155–80. Glenn Willumson's recent *Iron Muse: Photographing the Transcontinental Railroad* (Berkeley: University of California Press, 2013) is the newest take on Russell and how his photography was used to tell the story of the transcontinental railroad. See also Barry B. Combs, *Westward to Promontory: Building the Union Pacific across the Plains and Mountains: A Pictorial Documentary* (Palo Alton: American West Publishing, 1969); Anne Farrar Hyde, *An American Vision: Far Western Landscape and National Culture, 1820–1920* (New York: New York University Press, 1990), 93–96; and Shawn Michelle Smith, *At the Edge of Sight: Photography and the Unseen* (Durham, NC: Duke University Press, 2013), esp. chap. 4, "Preparing the Way for the Train: Andrew J. Russell," 99–127.

3 Weston J. Naef, and James N. Wood, *Era of Exploration: The Rise of Landscape Photography in the American West, 1860–1885* (New York: Albright-Knox Gallery and Metropolitan Museum of Art, 1975), 201; Sandweiss, *Print the Legend*, 77; Alan Trachtenberg, *Reading American Photographs: Images as History, Matthew Brady to Walker Evans* (New York: Hill and Wang, 1989), 107; "American Artists," *NN* 2, no. 12 (Dec. 15, 1860).

4 Tissandier, *History and Handbook of Photography*, 6–7, 34–36, 149–50.

5 Oliver Wendell Holmes, "The Stereoscope and the Stereograph," *AM* 3, no. 20 (June 1859): 748.

6 Alan Trachtenberg, "Photography: The Emergence of a Keyword," in *Photography in Nineteenth-Century America*, ed. Martha Sandweiss (Fort Worth, TX: Amon Center Museum, 1991), 26–27. The idea that one's image is material has illustrious forebears: Democritus, Epicurus, and Lucretius theorized that all bodies gave off essences of themselves. The insistence continues to this day. Roland Barthes wrote: "The photograph is literally an emanation of the referent. From a real body, which was there, proceed radiations which ultimately touch me, who am here." See Roland Barthes, *Camera Lucida*, trans. Richard Howard (New York: Hill and Wang, 1981), 81.

7   Ned Buntline, *Love at First Sight; or, The Daguerreotype. A Romantic Study of Real Life* (Boston: Lerow and Company, [1848?]), 9.
8   A. J. Russell shot on a number of differently sized glass negatives, but the most common for his single railroad views was ten inches by thirteen inches. Combs, *Westward to Promontory*, 18. William D. Pattison, "The Pacific Railroad Rediscovered," *Geographical Review* 52, no. 1 (Jan. 1962): 25–36; and Glenn Willumson, "'Photographing under Difficulties': Andrew Russell's Photographs for the King Survey," in *Framing the West: The Survey Photographs of Timothy H. O'Sullivan*, ed. Toby Jurovics, Carol M. Johnson, Glenn Willumson, and William F. Stapp (New Haven, CT: Yale University Press, 2010), 234n24.
9   W. H. Jackson, who was the official photographer for F. V. Hayden's 1870 Geological and Geographical Survey of the Territories, wrote of the tribulations of outdoor photography: "Working under a blazing sun, on a dry, parched, and dusty sage brush plain, my first taste of the realities of outdoor photography was not of the rose-water order." W. H. Jackson, "Field Work," *PP* 12, no. 135 (Mar. 1875): 92, 93; Beaumont Newhall, *The History of Photography from 1839 to the Present* (New York: Museum of Modern Art, 1982), 100–103.
10 See Joel Snyder, *One/Many: Western American Survey Photographs by Bell and O'Sullivan* (Chicago: David and Alfred Smart Museum of Art, 2006), 75–79.
11 This description of how one would prepare, make, and develop a glass-plate negative comes from Tissandier, *History and Handbook of Photography*, 114–17, 122–23, 126–28, 130–33; Holmes, "Stereoscope and the Stereograph," 740; and Oliver Wendell Holmes, "Doings of the Sunbeam," *AM* 12, no. 69 (July 1863): 12.
12 Jonathan Bordo has noted that figures in landscape images are often witnesses to the drawing of boundaries between wilderness and culture. He then goes on to suggest that, when humans are absent, trees often stand in for them. See Jonathan Bordo, "Picture and Witness at the Site of Wilderness," in *Landscape and Power*, 2nd ed., ed. W. J. T. Mitchell (Chicago: University of Chicago Press, 2002), 297, 299. Russell's photograph of the tree can be found in F. V. Hayden, *SPRM*; A. J. Russell, *PA*; A. J. Russell, *CUPR*; and as stereographic views. There are actually four different views of the tree taken by Russell, three of which, entitled *1000 Mile Tree, Looking South*; *1000 Mile Tree Gorge*, *1000 Mile Tree Excursion Tree Excursion Party*; and *1000 Mile Tree, Eastern Portal* can, as far as I know, be found only as stereo views. The other, *1000 Mile Tree, Excursion Party*, is the same as the image I reproduce here. My deep thanks to Glenn Willumson, who has very generously shared his research with me.
13 See James M. Reilly, *The Albumen & Salted Paper Book: The History and Practice of Photographic Printing, 1840–1895* (New York: Light Impressions Corporation, 1980).

14 Barbara Novak argues that images of stumps can paradoxically signify both Progress and a great wasting loss, an insight that Nicolai Cikovsky further develops by arguing that stumps and felled trees were two of the most common and widely understood iconographic devices in nineteenth-century image making. See Barbara Novak, "The Double-Edged Axe," *Art in America* 64 (Jan.-Feb. 1976): 45–50; and Nicolai Cikovsky Jr., "'The Ravages of the Axe': The Meaning of the Tree Stump in Nineteenth-Century American Art," *Art Bulletin* 61, no. 4 (Dec. 1979): 611.

15 New York Public Library, *Tracking the West: A. J. Russell's Photographs of the Union Pacific Railroad* (n.p.: 1994), 4.

16 Two other photographers also made views of the scene: Charles Savage and Alfred A. Hart. Because of the complexities of the ownership and attribution of authorship to nineteenth-century photos, Savage was often given credit for Russell's iconic photo—and sometimes still is. Russell wasn't rediscovered until the early 1960s. See Pattison, "Pacific Railroad Rediscovered," 35.

17 There's a large literature on the transcontinental railroad and its cultural, political, and economic effects. I've relied most heavily on Maury Kline, *Union Pacific* (Minneapolis: University of Minnesota Press, 1987); Sarah H. Gordon, *Passage to Union: How the Railroads Transformed American Life, 1829–1929* (Chicago: Ivan R. Dee, 1996), Robert V. Hine and John Mack Faragher, *The American West: A New Interpretive History* (New Haven, CT: Yale University Press, 2000); and Richard White, *Railroaded: The Transcontinentals and the Making of Modern America* (New York: W. W. Norton & Co., 2011).

18 Heather Cox Richardson, *West from Appomattox: The Reconstruction of America after the Civil War* (New Haven, CT: Yale University Press, 2007), 77.

19 A. J. Russell, "Correspondence: The Laying of the Last Rail on the Union Pacific Rail Road—Nunda Is Honored with a Representation—Some Important Statements from Capt. Andrew J. Russell of Nunda," *NN* 10, no. 22 (May 29, 1869).

20 Williams, "Truth Be Told," 38.

21 Holmes, "Stereoscope and the Stereograph," 747.

22 For histories detailing scientific objectivity and the camera, see Lorraine Daston and Peter Galison, "The Image of Objectivity," *Representations* 40 (1992): 81–128; and Jennifer Tucker, *Nature Exposed: Photography as Eyewitness in Victorian Science* (Baltimore: Johns Hopkins University Press, 2005).

23 Trachtenberg, *Reading American Photographs*, 27.

24 Many, including Oliver Wendell Holmes, saw this particular application of the photograph as part of the medium's utopian potential. See Holmes, "Doings of the Sunbeam," 11: Tissandier, *History and Handbook of Photography*, 327. See also Allan Sekula, "Body and the Archive," in *The Contest of Meaning: Critical Histories of Photography*, ed. Richard Bolton (Cambridge,

MA: MIT Press, 1989), 343, 345, passim. The role of the camera in surveillance has been the subject of an enormous body of scholarship, and includes Richard Bolton, ed., *The Contest of Meaning: Critical Histories of Photography* (Cambridge, MA: MIT Press, 1989); W. J. T. Mitchell, ed., *Landscape and Power*, 2nd ed. (Chicago: University of Chicago Press, 2002); Joan M. Schwartz and James R. Ryan, eds., *Picturing Place: Photography and the Geographical Imagination* (London: I. B. Tauris, 2003); and Smith, *Edge of Sight*.

25 Keith F. Davis, *The Origins of American Photography: From Daguerreotype to Dry Plate, 1839–1885* (New Haven, CT: Yale University Press, 2007), 49.

26 Barthes, *Camera Lucida*, 88–91, 96, 117. I once fell in love with a photograph. It was in an old German iron-cross cemetery, somewhere north of Minot, North Dakota. A woman in all her wedding finery, and who seemed to be my age when her picture was taken, stared out at me from the photographic likeness fixed to her monument. There was no name, and no possible chance of introduction. I was left with an aching sense of sadness.

27 Edgar Allen Poe, "The Daguerreotype," in *Classic Essays on Photography*, ed. Alan Trachtenberg (New Haven, CT: Leete's Island Books, 1980), 37.

28 Credit is often given to Asa Whitney as the first to dream up a transcontinental railroad, but his work began ten years after Plumbe's.

29 John Plumbe, *Memorial against Mr. Asa Whitney's Railroad Scheme* (Washington, DC: Buell & Blanchard, 1851), 20.

30 John Plumbe Jr., *Sketches of Iowa and Wisconsin, Taken during a Residence of Three Years in those Territories* (1839; repr., Iowa City: State Historical Society of Iowa, 1948), v, vi; and Clifford Krainik, "National Vision, Local Enterprise: John Plumbe Jr. and the Advent of Photography in Washington, D.C.," *Washington History: Magazine of the Washington Historical Society of Washington, D.C.*, Fall–Winter 1997–1998, 7.

31 Plumbe, *Memorial*, 32, passim.

32 Davis, *Origins of American Photography*, 24; and Krainik, "National Vision, Local Enterprise," 9.

33 Partly this was due to the quality of a Plumbe daguerreotype, and he won a number of medals and premiums for "the most beautiful colored Daguerreotypes and best Apparatus ever exhibited." But Plumbe's fame also rested on his business acumen: he pioneered the franchise, and opened international branches of the Plumbe National Daguerrian Gallery and Photographic Depots. He also pioneered brand-name appeal, having every daguerreotype impressed "PLUMBE." See John Doggett Jr., *The Great Metropolis; or, Guide to New York for 1846* (New York: H. Ludwig, 1846), 118; Krainik, "National Vision, Local Enterprise," 11, 15; and Barbara McCandless, "The Portrait Studio and the Celebrity: Promoting the Art," in *Photography in Nineteenth-Century America*, ed. Martha Sandweiss (Fort Worth, TX: Amon Center Museum, 1991), 52.

34 "A Visit to Plumbe's Gallery," *BDE* 5, 160 (July 2, 1846). Though anonymous,

this article is often credited to Walt Whitman, in whose life Plumbe would play no small part. The daguerreotypist who took the image of Whitman as a young tough—a lithograph of which appears on the flyleaf of the 1855 edition of *Leaves of Grass*—was trained by Plumbe, and the ability to reproduce a daguerreotype through lithography was first pioneered by Plumbe as well. See Reynolds, *Walt Whitman's America*, 280–82, 284.

35 John R. Plumbe, *The Settlers and Land Speculators of Sacramento* (New York: n.p., 1851), 7, 13.
36 Plumbe, *Memorial*, 2–3, 23.
37 "Visit to Plumbe's Gallery," *BDE*.
38 Russell, "On the Mountains with the Tripod and Camera," 34.
39 Newberry envisioned a time when "the cities that now stand upon [the Great Lakes'] shores will . . . have grown colossal in size, then gray with age, then have fallen into decadence and their sites be long forgotten, but in the sediments that are now accumulating in these lake-basins will lie many a wreck and skeleton, tree-trunk and floated leaf. Near the city sites and old river mouths these sediments will be full of relics that will illustrate and explain the mingled comedy and tragedy of human life." Hayden, *Sun Pictures*, 150. For a fantastic discussion of how Russell's images were used by the railroad and the popular press, see Willumson, *Iron Muse*.
40 Besides those sources already mentioned, Russell's photos also appeared in A. J. Russell, *GWI*; A. J. Russell, *UPRRV*; *GWW*; and later as a series of stereographic cards titled *Union Pacific R.R. Stereoscopic Views and Pacific R.R. Views across the Continent West from Omaha* (often credited to O. C. Smith, who obtained the negatives in 1875 and published them under his own name); S. J. Sedgwick, *Announcement of Prof. S. J. Sedgwick's Illustrated Course of Lectures and Catalogue of Stereoscopic Views*; William D. Pattison, "Westward by Rail with Professor Sedgwick: A Lantern Journey," *Historical Society of Southern California Quarterly* 42, no. 4 (Dec. 1960): 336; Holmes, "Stereoscope and the Stereograph," 747; John S. Waldsmith, *Stereo Views: An Illustrated History and Price Guide* (Radnor, PA: Wallace-Homestead Book Company, 1991), 51–52. The best information on Russell's railroad stereo views can be found online at the Central Pacific Railroad Photographic History Museum, at http://cprr.org/Museum/Russell_Catalog.html (accessed May 20, 2011).
41 This is the argument arrived at by two of the historians most critically invested in Russell's work, Martha Sandweiss and Susan Danly, each of whom focuses her analysis not on the images themselves but on a preexisting hegemonic ideology of "the culture industry." As Sandweiss puts it, "The public photographs that are the focus of this book [*Print the Legend*] were designed for widespread consumption, and intended to pass before the eyes and into the hands of strangers." But Sandweiss adds another step to her analysis: we know that Russell's photos were used to sell something because

that's what the captions, titles, and surrounding text do; and for Sandweiss, text is the only way to give an image meaning. Even in those rare cases when Sandweiss allows that photos convey information on their own, it is only through their sequencing in an album, which, Sandweiss argues, must be read from left to right, forward to back—that is, we must read collections of photos according to the logic of the printed word. For Danly, Russell's photography can be understood primarily as an example of a picturesque visual aesthetic, an aesthetic that she argues is inherently problematic in its easy accommodation of nature and industry. Like Sandweiss's, Danly's interpretation is based on things primarily beyond the photo's frame—although she does note at one point that there is a great deal of tension, in places, between Russell's photographs and the captions and literary text surrounding *Great West Illustrated*. See Sandweiss, *Print the Legend*, 4–8, 13, and esp. chap. 5, "'Westward the Course of Empire,'" and chap. 6, "'Momentoes of the Race,'" 155–273; Martha A. Sandweiss, "Dry Light: Photographic Books and the Arid West," in *Perpetual Mirage: Photographic Narratives of the Desert West*, ed. May Castleberry (New York: Whitney Museum of American Art, 1996), 24, 25–26; and Martha Sandweiss, "Undecisive Moments: The Narrative Tradition in Western Photography," in *Photography in Nineteenth-Century America*, ed. Martha Sandweiss (Fort Worth, TX: Amon Center Museum, 1991), 99, passim; Danly Walther, "Landscape Photographs of Alexander Gardner and Andrew Joseph Russell," 79–80; Danly, introduction to *Railroad in American Art*, 31; Susan Danly, "Photography, Railroads, and Natural Resources in the Arid West: Photographs by Alexander Gardner and A. J. Russell," in *Perpetual Mirage: Photographic Narratives of the Desert West*, ed. May Castleberry (New York: Whitney Museum of American Art, 1996), 51, 55; Horkheimer and Adorno, *Dialectic of Enlightenment*, 95.

42 The notion of a visual economy has gained traction recently, and it is how Willumson approaches his study of Russell. In what he calls "the social production of the original images," Willumson argues that meaning is made when the photographer makes his view; when the employer exercises his corporate authority in selecting, printing, captioning, advertising, and selling the image; and when the press uses the image for its own purposes. Willumson, *Iron Muse*, 3–9. And so, though an image's meaning might not exist independently from literary text, images do bring something of their own to the interpretive table. A visual economics opens up logics of reading images that gesture beyond that of the book: for instance, stereographic sets were often broken up, interrupting a front-to-back, left-to-right organization—and so one could trace the circulation of whole or partial sets of images, or even individual images as they appear in the archive. For other examples of how one might practice a visual economy, see Richard Bolton, introduction to *The Contest of Meaning: Critical Histories of Photography*, ed. Richard Bolton (Cambridge, MA: MIT Press, 1989); Elizabeth Edwards,

Raw Histories: Photographs, Anthropology and Museum (Oxford, UK: Berg, 2001); Schwartz and Ryan, *Picturing Place*; Finis Dunaway, *Natural Visions: The Power of Images in American Environmental Reform* (Chicago: University of Chicago Press, 2005); and Errol Morris, *Believing Is Seeing (Observations on the Mysteries of Photography)* (New York: Penguin Press, 2011).

43 Willumson, *Iron Muse*, 167.

44 Walter Benjamin, "Brief History of Photography," in *One Way Street and Other Writings*, trans. J. A. Underwood (New York: Penguin Books, 2009), 184.

45 The idea that images are alive, as well as the notion of a critical idolatry, all come from W. J. T. Mitchell, whose trilogy—*Iconology: Image, Text, Ideology*, *Picture Theory: Essays on Verbal and Visual Representation*, and *What Do Pictures Want? The Lives and Loves of Images*—is the basis for my own practice of visual ecology. Mitchell's work is dedicated to the proposition that images should bear equal epistemological weight to text. Debates over the hierarchy of image and text, Mitchell argues, are ultimately debates favoring a dualistic understanding of the world, between "body and soul, world and mind, nature and culture"; or, in other words, one of the battlegrounds of modernity. That is, there's a politics to interpretation, and I've found the work of John Berger both delightful and intellectually sustaining on this front. "It might be possible," Berger wrote in *About Looking*, "to begin to use photographs according to a practice addressed to an alternative future. This future is a hope which we need now, if we are to maintain a struggle, a resistance, against the societies and culture of capitalism." We can do this, Berger argued, by reanimating photographs, reinserting them into something of their original contingency, a "living context," so that "they would continue to exist in time, instead of being arrested moments." What he started to sketch out is the rudiments of a visual ecology, one built by "constructing a context for a photograph . . . construct it with words . . . construct it with other photographs . . . construct it by its place in an ongoing text of photographs and images." Finally, my thinking on visual ecology owes a great deal to Susan Sontag's provocative, and bleak, *On Photography* (to which Berger was responding). It's worth quoting her at length: "A capitalist society requires a culture based on images. It needs to furnish vast amounts of entertainment in order to stimulate buying and anesthetize the injuries of class, race, and sex. And it needs to gather unlimited amounts of information, the better to exploit natural resources, increase productivity, keep order, make war, give jobs to bureaucrats. The camera's twin capacities, to subjectivize reality and to objectify it, ideally serve these needs and strengthen them. Cameras define reality in two ways essential to the workings of an advanced industrial society: as a spectacle (for masses) and as an object of surveillance (for rulers)."

Sontag, to her great credit, does leave one door open—the ecological door: "If there is a better way for the real world to include the one of images, it will require an ecology not only of real things, but of images as well."

Finally, theorizing images as ecological members carries with it the methodological possibility of following images according to their own visual logic, rather than according to the logic of text. There's a sort of intervisuality that allows for a photograph to ecologically extend beyond its frame to other images and visual conventions. This is partly what makes photographs expressive: they gesture beyond the confines of their borders, calling other images to mind. See W. J. T. Mitchell, *Iconology: Image, Text, Ideology* (Chicago: University of Chicago Press, 1986), 41, 46, 49; W. J. T. Mitchell, *Picture Theory: Essays on Verbal and Visual Representation* (Chicago: University of Chicago Press, 1994), esp. chap. 3, "Beyond Comparison: Picture, Text, and Method," 83–107; W. J. T. Mitchell, *What Do Pictures Want: The Lives and Loves of Images* (Chicago: University of Chicago Press, 2005), 27, 47–48; John Berger, *About Looking* (New York: Vintage International, 1980), 55–56, 60–67; and Susan Sontag, *On Photography* (New York: Picador, 1977), 5, 178–79, 180. For something like a visual ecology of photojournalism, see Robert Hariman and John Luis Lucaites, *No Caption Needed: Iconic Photographs, Public Culture, and Liberal Democracy* (Chicago: University of Chicago Press, 2007). For a delightful praxis of visual ecology, see Lawrence Weschler, *Everything That Rises: A Book of Convergences* (San Francisco: McSweeney's, 2006). And for a history exploring why critics and scholars have come to distrust vision, see Martin Jay, *Downcast Eyes: The Denigration of Vision in Twentieth-Century French Thought* (Berkeley: University of California Press, 1993).

46 A. J. Russell, "Photographic Reminiscences of the Late War," *APB* 13 (1882): 212.
47 Russell, "Photographic Reminiscences of the Late War," 212.
48 "Panorama of the War," *NN* 3, no. 11 (Dec. 14, 1861).
49 Williams, "'Richmond Again Taken,'" 437–38.
50 Herman Haupt, *Reminiscences of General Herman Haupt* (Milwaukee, WI: Wright & Joys Co., 1901), 256; Charles F. Cooney, "Andrew J. Russell: The Army's Forgotten Photographer," *Civil War Times Illustrated* 21 (Apr. 1982): 33; and Keith F. Davis, "'A Terrible Directness: Photography of the Civil War Era,'" in *Photography in Nineteenth-Century America*, ed. Martha Sandweiss (Fort Worth, TX: Amon Center Museum, 1991), 157.
51 New York Public Library, *Tracking the West*, 4; Williams, "Richmond Again Taken," 438, 442–45; Hyde, *American Vision*, 108.
52 "Panorama," *NN*.
53 "Capt. Russell Photographing Battle Scenes," *NN* 5, no. 37 (June 13, 1863).
54 Russell, "Photographic Reminiscences of the Late War," 212, 213.
55 This desire and power of the photograph is something Roland Barthes calls "punctum." It is this element that "rises from the scene, shoots out of it like

an arrow, and pierces me." Barthes, *Camera Lucida*, 26–27; John Berger and Jean Mohr, *Another Way of Telling* (New York: Vintage International, 1982) 117–18, 128–29.

56 Russell loved to tell stories, with himself as the hapless protagonist, and many of them have to deal with the recalcitrance of mules. "New York Correspondence," *PP* 7 (1870): 82; A. J. Russell, "Letter from Mormondom," *NN* 10, no. 34 (Aug. 21, 1869).

57 *Andrew J. Russell: Visual Historian.*

58 For a brilliant investigation of how landscape artists hired for western surveys negotiated between their need for artistic expression and the expectations of their government patrons, see Robin Kelsey's *Archive Style: Photographs & Illustrations for U.S. Surveys, 1850–1890* (Berkeley: University of California Press, 2007).

59 John Conron, *American Picturesque* (University Park: Pennsylvania State University, 2000), esp. chap. 2, "Eclecticism and the Multiplication of Feelings in Form," 17–32; Andrew Wilton and Tim Barringer, *American Sublime: Landscape Painting in the United States, 1820–1880* (London: Tate Publishing, 2002), 13; and Marjorie Hope Nicolson, *Mountain Gloom and Mountain Glory: The Development of the Aesthetics of the Infinite* (1959; repr., Seattle: University of Washington Press, 1997), 299.

60 Snyder points out that many western photos made by Timothy H. O'Sullivan and William Bell are "roadblocks" in that they don't strictly conform to the picturesque, sublime, or beautiful. Similarly, Robin Kelsey argues that "those responsible for producing survey pictures did not encounter pictorial conventions as strict and precise prescriptions." Snyder, *One/Many*, 27–28; and Kelsey, *Archive Style*, 16.

61 As Susan Sontag noted, aesthetics of beauty are always "a quasi-moral project." Susan Sontag, *At the Same Time: Essays and Speeches*, ed. Paolo Dilonardo and Anne Jump (New York: Farrar, Straus & Giroux, 2007), 11.

62 Edmund Burke, *A Philosophical Enquiry into the Origins of our Ideas of the Sublime and Beautiful*, ed. Adam Phillips (1757; repr., Oxford: Oxford University Press, 1990), 107.

63 Nicolson, *Mountain Gloom and Mountain Glory*, esp. chap. 3, "New Philosophy," and chap. 4, "The Geological Dilemma," 113–83; Wilton and Barringer, *American Sublime*, 14–21; David E. Nye, *American Technological Sublime* (Cambridge, MA: MIT Press, 1994), 4–6; and Mitchell, *Iconology*, 125–29.

64 Burke, *Philosophical Enquiry*, 36.

65 An important twist on the classic Burkean sublime can be seen in a good deal of American landscape painting—think of Leutze's *Westward the Course of Empire Takes Its Way*. Because the sublime had once been associated with the grand acts of history, there's a way that the course of human history can be rendered sublime when that history is grand enough, such as in the "conquering" of a continent. Burke, *Philosophical Inquiry*, 79–83.

66 Nicolson, *Mountain Gloom and Mountain Glory*, 299, 323.
67 David Marshall has provocatively argued that the picturesque, rather than distancing the viewer from the land, exists to productively blur the distinctions between human and nature, representation and reality, seer and seen. David Marshall, *The Frame of Art: Fictions of Aesthetic Experience, 1750–1815* (Baltimore: Johns Hopkins University Press, 2005), esp. chap. 1, "The Problem of the Picturesque," 16–39.
68 William Gilpin, *Remarks on Forest Scenery and Other Woodland Views*, ed. Sir Thomas Dick Lauder (Edinburgh: Fraser & Co., 1834), 1:45, 47, 49–53; Sir Uvedale Price, *On the Picturesque: With an Essay on the Origin of Taste, and Much Original Matter*, ed. Sir Thomas Dick Lauder (Edinburgh: Caldwell, Lloyd, and Co., 1842), 60–62.
69 Burke, *Philosophical Enquiry*, 83.
70 Price, *On the Picturesque*, 504–10; Nicolson, *Mountain Gloom and Mountain Glory*, 16, 271, 281–82, 287.
71 Burke, *Philosophical Enquiry*, 47; Susan Sontag and Elaine Scarry would agree. See Sontag, *At the Same Time*, 13; and Elaine Scarry, *On Beauty and Being Just* (Princeton, NJ: Princeton University Press, 1999), 90.
72 Thomas Cole, "Essay on American Scenery," in *Thomas Cole: The Collected Essays and Prose Sketches*, ed. Marshall Tymn (St. Paul's: John Colet Press, 1980), 4.
73 See Nye, *American Technological Sublime*.
74 Deborah Bright sees landscape as a bastion of American myths about nature, culture, beauty, and masculinity, and Barbara Novak argues that in nineteenth-century American art, the sublime was grafted onto nationalism, "under which the aggressive conquest of the country could be accomplished." This is part of the sublime tradition that Ansel Adams comes out of, the celebration of American Nature as a stand-in for (manly) American history. The photographer and essayist Robert Adams has a wonderful, nuanced, sensitive critique of Adams—one that calls attention to the beauty and artistry of Adams's work even as it refuses to let Adams off the hook. "Like many of his generation," Robert Adams writes, "he seems never really to have faced what his country's economic system meant for the land." Deborah Bright, "Of Mother Nature and Marlboro Men: An Inquiry into the Cultural Meanings of Landscape Photography," in *The Contest of Meaning: Critical Histories of Photography*, ed. Richard Bolton (Cambridge, MA: MIT Press, 1989); Barbara Novak, *Nature and Culture: American Landscape and Painting, 1825–1875*, rev. ed. (New York: Oxford University Press, 1995), 34–38; Robert Adams, *Why People Photograph: Selected Essays and Reviews* (New York: Aperture, 1994), 114.
75 Barbara Novak argues convincingly: "The loss of pleasure is not a sidebar to the often grim interrogations of artworks' hidden agendas. What does it mean when such words as 'pleasure' and its synonyms are denied entry

into the discourse, indeed proscribed? . . . The denial of pleasure, however, is itself suppression of an intrinsic component of [art's] perception." More recently, theorist Elaine Scarry has linked beauty to justice, and at times sounds like Thomas Cole and Edmund Burke: "Beauty," she writes, "really is allied with truth," for it gives us "the experience of conviction and the experience, as well, of error." To lose beauty is to lose a generative liveliness and a politics of justice: "Beauty is . . . a compact, or a contract between the beautiful being . . . and the perceiver. As the beautiful being confers on the perceiver the gift of life, so the perceiver confers on the beautiful being the gift of life." Novak, *Nature and Culture*, viii–x; and Scarry, *On Beauty and Being Just*, 52, 90. See also Felski, *Limits of Critique*.

76 Writing of the pastoral, Lawrence Buell notes that aesthetics cannot "be pinned to a single ideological position." And one of the great contributions of Finis Dunaway's work on American environmental image making is his argument that the sublime has a history, that it has changed over time—though he is primarily concerned with aesthetics in the twentieth century. Buell, *Environmental Imagination*, 44; Dunaway, *Natural Visions*, xix–xx, and esp. the epilogue, "The Ecological Sublime," 194–212.

77 The great promise of art is that it can help us to immerse ourselves to an extent that we can be overcome by wonder, as Jane Bennett puts it, and start to reimagine ourselves as part of a wider community, a community toward which we owe a great deal of responsibility. See Buell, *Environmental Imagination*, 77, 206, 219, 266, 425n1; and Bennett, *Enchantment of Modern Life*.

78 Van Noy, *Surveying the Interior*, 6–7, 33, 35.

79 Russell, "Letter from Mormondom."

80 Gilpin, *Remarks on Forest Scenery*, 45–48.

81 Ibid., 362–63.

82 Cole, "Essay on American Scenery," 5, 6. This is, I think, why Elaine Scarry is so keen to root good education in the appreciation of beauty—beauty forces us out of our own provincialism. Scarry, *On Beauty*, especially pt. 1, "On Beauty and Being Wrong," 1–53.

83 Walther, "Landscape Photographs of Alexander Gardner and Andrew Joseph Russell," 65–66.

84 No one knows quite how Russell ended up working for the Union Pacific, although Willumson notes that the corporation was filled with former Civil War veterans, including generals Grenville Dodge and Jack Casement, respectively the chief engineer and the lead track-laying contractor for the railroad. Dodge had been with the military railroad when Russell was shooting his photos. Willumson, *Iron Muse*, 48–49.

85 Ibid., 96.

86 Pattison and Smith argue for the $75 figure, while Danly and Willumson note $50. In any case, it was expensive. Pattison, "The Pacific Railroad Rediscov-

ered," 33n14; Smith, *At the Edge of Sight*, 110; Danly, *Andrew Joseph Russell's "The Great West Illustrated,"* 94; and Willumson, *Iron Muse*, 104.

87 Russell, *Great West Illustrated*. There are no page numbers in this text. The quotations can be found on the title page and in the preface. Subsequent citations are to plate numbers.

88 Barbara Novak has argued that the lone figure can either invite us into the frame or shut us out by keeping us from seeing what it sees. And though, overwhelmingly, Novak wants to incorporate mid-nineteenth-century American art into a culture of optimism, she does concede that these figures can seem like "intruders in the Garden of Creation." See Novak, *Nature and Culture*, 198, and more generally chap. 8, "Man's Traces: Axe Train, Figure," 157–200.

89 It is not at all clear who wrote the text for *Great West Illustrated*; most have credited Russell, given that his name appears on the title page; yet Willumson argues that the prose was most likely penned collaboratively, with someone at the UPRR. However, given the lowly status of photographers at the time, what seems like an uneasy working relationship between Russell and the UPRR, and the narrative disjunction between text and image (Willumson even notes that, though the prose in each copy of the album remains the same, the images do not, suggesting that "it was the selection of locations and the annotated captions, rather than specific photographs, that was of primary importance"), I doubt that the prose in *Great West Illustrated* gives us much of a sense of Russell's landscape vision. Willumson, *Iron Muse*, 97.

90 Russell, *Great West Illustrated*.

91 On catastrophism and the photographic aesthetics of Timothy H. O'Sullivan's work for Clarence King's Geological Exploration of the Fortieth Parallel survey, see Aaron Sachs, *The Humboldt Current: Nineteenth-Century Exploration and the Roots of American Environmentalism* (New York: Viking, 2006), 246–51. Oddly enough, in the summer of 1869, Russell bumped into Clarence King; although O'Sullivan was the official photographer for the expedition, Russell joined it for three weeks and made "a great many magnificent views in this vicinity," all of which, until very recently, have been credited to O'Sullivan. A. J. Russell, "Rocky Mountain Adventure," *NN* 10, no. 27 (Sept. 25, 1869). For a reattribution of Russell's photos initially credited to O'Sullivan, see Willumson, "'Photographing under Difficulties,'" 175–85.

92 On irony and landscape, see Jonathan Smith, "The Lie That Binds: Destabilizing the Text of Landscape," in *Place/Culture/Representation*, ed. James Duncan and David Ley (London: Routledge, 1993), 78–94.

93 For a reading of the iconology and iconography of the railroad in the American landscape, see Leo Marx, "The Railroad-in-the-Landscape: An Iconological Reading of a Theme in American Art," in *The Railroad in American Art: Representations of Technological Change*, ed. Susan Danly and Leo Marx (Cambridge, MA: MIT Press, 1988).

94 "The visual corollary of silence," writes Barbara Novak, "is stillness. Stillness and silence are the antipodes of the progressive noise and action of civilization." It's this loss of stillness that Robert Adams captures so well in his essay "In the Nineteenth-Century West." No longer is western space in the United States characterized by silence, resistance to speed, revelation by light. Perhaps the greatest triumph of the nineteenth-century photographers, and our greatest cultural loss, was their insistence that space, "the vacant center," as Adams put it, was an integral part of the coherence of the scene. If nature abhors a vacuum, the culture of capitalism cannot tolerate an untapped resource. Novak, "Double-Edged Axe," 48; Adams, *Why People Photograph*, 134, 138, 147.

95 Marx, "Railroad-in-the-Landscape," 206.

96 On an aggressive sublime as applied to the American West, see William H. Truettner, "Ideology and Image: Justifying Westward Expansion," in *The West as America: Reinterpreting Images of the Frontier, 1820–1920* (Washington, DC: Smithsonian Institution Press, 1991), 35.

97 The locomotive was the perfect symbol of modernity, Leo Marx argues in his analysis of John Stuart Mill's review of Tocqueville's *Democracy in America*, because the "superiority of the present to the past" was inherent in the engine's very form. Leo Marx, *The Machine in the Garden: Technology and the Pastoral Ideal in America* (London: Oxford University Press, 1976), 192.

98 Russell was not alone in showing the horrors of mining in the West: Timothy H. O'Sullivan captured cave-ins in 1867 and 1868. See Kelsey, *Archive Style*, 117–31; and Sachs, *The Humboldt Current*, 215–28.

99 Willumson, *Iron Muse*, 105.

100 Both Willumson and Smith note how strange it is that the train is nearly always absent from theses images, which are often taken to be celebrations of the railroad. Willumson, *Iron Muse*, 98; Smith, *At the Edge of Sight*, 105.

101 "Real art has the capacity to make us nervous," writes Susan Sontag. "By reducing the work of art to its content [by which Sontag means its underlying ideology] and then interpreting that, one tames the work of art. Interpretation makes art manageable, comfortable." Susan Sontag, "Against Interpretation," in *Against Interpretation and Other Essays* (New York: Picador, 1966), 8.

102 Sandweiss argues that Humphrey Lloyd Hine, the Canadian photographer who made a photo of a skull bleaching on the wide Canadian prairie in 1858, was vocally critiquing territorial expansion in his photo. Although she goes on to say that Hine's photo is "utterly unlike any photograph in the arid American West during the mid- to late-nineteenth century," Russell is clearly the exception—and so I think that he, too, deserves to be seen as critical of technological, territorial, and economic progress. Sandweiss, "Dry Light," 21.

103 "Annexation," *USM*, July–Aug. 1845, 5–10.

104 Nathaniel Hawthorne, "The Celestial Railroad," in *Hawthorne's Short Stories*, ed. Newton Arvin (New York: Vintage Classics, 2011), 237, 250.
105 Henry David Thoreau, *Walden; or, Life in the Woods*, in *WWMC*, 396; Henry George, "What the Railroad Will Bring Us," *Overland Monthly and Out West Magazine* 1, no. 4 (Oct. 1868): 302.
106 George, "What the Railroad Will Bring Us," 303.
107 For a discussion of how the earlier survey artist, Arthur Schott, likewise filled his images with ambivalent crosses, see Kelsey, *Archive Style*, 53–60.
108 "From Georgia," *Daily Cleveland Herald*, Mar. 28, 1866; *Norwich (CT) Aurora*, Oct. 27, 1869; "Washington," *Boston Post*, June 28, 1866; and "The Mormons," *Evening Post* (New York), June 5, 1866.
109 Matthew 27:33 (King James Version).
110 Barthes, *Camera Lucida*, 110, 115.
111 John Berger and Jean Mohr write that "every photograph presents us with two messages: a message concerning the event photographed and another concerning a shock of discontinuity." There's a way, they continue, that photographs are objective in that they can show only what was actually there. "Photographs do not translate from experience. They quote from them," write Berger and Mohr, in a Barthesian vein. Yet, what makes photographs soulful is not their indexical recording of facts: it is the ambiguity of these facts, the gap between fact and meaning. And in that gap lies human astonishment. Ambiguity, then, may be more objective, a better way to get at what things were *like* than a clear-cut, empirical method. "Photographs [when placed in a sequence] are restored to a living context . . . a context of experience. And there, their ambiguity at last becomes true. It allows what they show to be appropriated by reflection. The world they reveal, frozen, becomes tractable. The information they contain becomes permeated by feeling. Appearances become the language of a lived life." We make meaning out of photos in the same way that we make meaning out of the world around us, in the practice of our everyday lives. See Berger and Mohr, *Another Way of Telling*, esp. "Appearances," 82–129, and "Stories," 279–89. Quotations on 86, 96, 289. Trachtenberg, *Reading American Photographs*, xvi; and Edwards, *Raw Histories*, 3, 11, 22.
112 Shawn Michelle Smith, channeling Benjamin's notion of an optical unconscious, notes that Russell was photographing what would soon disappear in the Iron Horse's blur of speed. Benjamin, "Brief History of Photography," 176; and Smith, *At the Edge of Sight*, 4–6, 100.
113 There are at least a small handful of Russell's photographs that capture these wasteful efforts, though they are never the primary focus of his camera's lens. For an example, see Combs, *Westward to Promontory*, 67. For the railroads' political influence, see Hine and Faragher, *American West*, 218–82; Thomas Weston Fels, *Destruction and Destiny: The Photographs of A.J. Russell; Directing American Energy in War and Peace, 1862–1869* (Pittsfield, MA:

Berkshire Museum, 1987), 14. For a detailed examination of the transcontinentals' vast fleecing of the public, see White, *Railroaded*.

114 One of the reasons that photos are so powerful is that they are indexes of a real, unique event, a fingerprint. As Sontag put it, "A photograph is not only an image (as a painting is an image), an interpretation of the real; it is also a trace, something directly stenciled off the real, like a footprint or a death mask." Sontag, *On Photography*, 154.

115 "Thus, [photographs] trade simultaneously on the prestige of art and the magic of the real. They are clouds of fantasy and pellets of information." Ibid., 69.

116 Leo Marx, "Technology: The Emergence of a Hazardous Concept," *Social Research* 64, no. 3 (Fall 1997): 978.

117 Susan Danly, introduction to *The Railroad in American Art: Representations of Technological Change*, ed. Susan Danly and Leo Marx (Cambridge, MA: MIT Press, 1988), 6, 17, 31.

118 John Muir, *Steep Trails*, in *John Muir: The Eight Wilderness Discovery Books* (Seattle: Mountaineers, 1992), 1000.

119 Russell, "Laying of the Last Rail."

120 Ibid.

121 Drew Gilpin Faust, *This Republic of Suffering: Death and the American Civil War* (New York: Vintage Books, 2008), xiv, 60, 210; Megan Kate Nelson, *Ruin Nation: Destruction and the American Civil War* (Athens: University of Georgia Press, 2012); Aaron Sachs, *Arcadian America: The Death and Life of an Environmental Tradition* (New Haven, CT: Yale University Press, 2013) esp. chap. 4, "Stumps," 137–209.

122 Gordon, *Passage to Union*, 144–48, 150, 152–53; Trachtenberg, *Incorporation of America*, 58; David Nye, "Visualizing Eternity: Photographic Constructions of the Grand Canyon," in *Picturing Place: Photography and the Geographical Imagination*, ed. Joan M. Schwartz and James R. Ryan (London: I. B. Tauris, 2003), 76.

123 Richardson, *West from Appomattox*, 342.

124 Marx, *Machine in the Garden*, 194; Alexander Pope, *Peri Bathous: or, Martinus Scriblerus; His Treatise of the Art of Sinking in Poetry*, in *The Prose Works of Alexander Pope*, ed. Rosemary Cowler (Hamden, CT: Archon Books, 1986), 2:211.

125 The phrase "beyond wilderness" comes from an issue of the photo-criticism journal *Aperture*, which, in 1990, devoted an entire issue to landscape aesthetics that reached beyond an Ansel Adams–esque wilderness vision. As Barry Lopez put it in his lead essay, "Unbounded Wilderness," "The fate of the American landscape . . . cannot be addressed solely in terms of 'wilderness' or be solved by 'wilderness preservation.'" Wilderness, he goes on to say, "hurts us in two ways. It preserves a misleading and artificial distinction between 'holy' and 'profane' lands, and it continues to serve the

industries that most seriously threaten wilderness." The solution, Lopez writes, is to rethink an ethic guided by an "informed reciprocity," an ethic that landscape photographers can help to envision, one that incorporates "not only our farmsteads and the retreats of the wolverine but the land upon which our houses, our stores, and our buildings stand. Our behavior, from planting a garden to mining iron ore, must begin to reflect the same principles." It's a vision that focuses not on replacing one ahistorical category, "wilderness," with another, "home," but on a moving, flexible "commensal relationship," one that can recognize things like class and exploitation. See Barry Lopez, "Unbounded Wilderness," *Aperture* 120 (Late Summer 1990): 2, 14.

126 Only recently, in the emergence of a school known as "New Topographics," have landscape photographs been made that resonate with Russell's work (although many practitioners of New Topographics might disavow this connection). See Robert Adams, Lewis Baltz, Bernd Becher and Hilla Becher, Joe Deal, Frank Gohlke, Nicholas Nixon, John Schott, Stephen Shore, and Henry Wessell Jr., *New Topographics* (Göttingen, Germany: Steidl Publishers, 2010); Mark Klett, Ellen Manchester, JoAnn Verburg, Gordon Bushaw, Rick Dingus, and Paul Berger, *Second View: The Rephotographic Survey Project* (Albuquerque: University of New Mexico Press, 1984), 5, 9; and Mark Ruwedel, *Westward the Course of Empire* (New Haven, CT: Yale University Art Gallery, 2008). For a historical framing of New Topographics, see Finis Dunaway, "Beyond Wilderness: Robert Adams, New Topographics, and the Aesthetics of Ecological Citizenship," in *Reframing the New Topographics*, ed. Greg Foster-Rice and John Rohrbach (Chicago: Center for American Places, Columbia College Chicago, 2010).

127 Thomas Stevens, "Across America on a Bicycle: III—Through Mormon Land, Over the Rocky Mountains, and on the Great Plains," *Outing: An Illustrated Monthly Magazine of Recreation* 6, no. 3 (June 1885): 290.

128 "An Outing," *Salt Lake (UT) Herald*, Sept. 23, 1888.

ACT FOUR

1   David Mason, *Ludlow* (Pasadena, CA: Red Hen Press, 2007), 101.
2   As with any superlative, there are qualifications: there are taller trees, trees with bigger girths, trees whose root system covers a larger area. But in terms of sheer wood volume, General Sherman stands a head above the rest.
3   Albert D. Richardson, *Beyond the Mississippi: From the Great River to the Great Ocean* (Hartford, CT: American Publishing, 1867), i.
4   Fitz-Hugh Ludlow, "Seven Weeks in the Great Yo-Semite," *AM* 13, no. 80 (June 1864): 740.
5   Karl Marx, "The Global Consequences of the Discovery of Gold in Califor-

nia," in *The Karl Marx Library*, vol. 2, *On America and the Civil War* (New York: McGraw-Hill, 1972), 14.

6   Surprisingly, there is very little humanistic scholarly literature concerned with the sequoias. The most important for my work include Simon Schama, *Landscape and Memory* (New York: Vintage Books, 1995); Lori Vermaas, *Sequoia: The Heralded Tree in American Art and Culture* (Washington, DC: Smithsonian Books, 2003); Walter Fry and John R. White, *Big Trees* (Stanford, CA: Stanford University Press, 1938); Larry M. Dilsaver and William C. Tweed, *Challenge of the Big Trees: A Resources History of Sequoia and Kings Canyon National Parks* (Three Rivers, CA: Sequoia Natural History Association, 1990); Donald Culross Peattie, *A Natural History of North American Trees* (1948; repr., Boston: Houghton Mifflin Company, 2007); and Jared Farmer's *Trees in Paradise: A California History* (New York: W. W. Norton & Co., 2013), esp. chap. 1, "Twilight of the Giants," 7–59. Of these, Vermaas's *Sequoia* is the only book-length treatment of the cultural reception and incorporation of the trees.

7   Peattie, *Natural History of North American Trees*, 18.

8   Shirley Sargent, ed., *Seeking the Elephant, 1849: James Mason Hutchings' Journal of His Overland Trek to California Including his Voyage to America, 1848 and Letters from the Mother Lode* (Glendale, CA: Arthur H. Clark Company, 1980), 22–24.

9   J. M. Hutchings, *In the Heart of the Sierras* (Oakland, CA: Pacific Press Publishing House, 1886), 214–15. Hutchings was a tireless and influential promoter of the Sierras and Big Trees. It is his version of their discovery that has become canonical.

10 In 1833, Captain Joseph R. Walker's party stumbled upon the trees, and one of the soldiers, Zenas Leonard, even published an account in 1839. Then, in 1841, John Bidwell saw the trees, followed by prospector J. M. Wooster in 1850. But Dowd's news was the first to gain widespread attention. Fry and White, *Big Trees*, 8–10; and Francis P. Farquhar, "Exploration of the Sierra Nevada," *California Historical Society Quarterly* 4, no. 1 (Mar. 1925): 6.

11 Hutchings, *In the Heart of the Sierras*, 219.

12 As with so much of the lore surrounding the Big Trees, this story is told in multiple places. I've stitched together this narrative from Geological Survey of California, *The Yosemite Book; A Description of the Yosemite Valley and the Adjacent Region of the Sierra Nevada, and of the Big Trees of California, Illustrated by Maps and Photographs* (New York: Julius Bien, 1868), 107; Thomas Starr King, *A Vacation among the Sierras*, ed. John A. Hussey (San Francisco: Book Club of California, 1962) 31; "Mammoth Tree from California," in *Giants in the Earth: The California Redwoods*, ed. Peter Johnstone (Berkeley: Heyday Books, 2001), 85–86; Schama, *Landscape and Memory*, 186–87; and Peattie, *Natural History of North American Trees*, 10.

13 For more on Hutchings, see Jen A. Huntley's nuanced study, *The Making of*

Yosemite: James Mason Hutchings and the Origins of America's Most Popular National Park (Lawrence: University Press of Kansas).

14 Ludlow, "Seven Weeks in the Great Yo-Semite," 740; Witold Rybczynski, *A Clearing in the Distance: Frederick Law Olmsted and America in the 19th Century* (New York: Simon & Schuster), 223.

15 King, *Vacation among the Sierras*, ix, 16, 17.

16 A. L. Kroeber, *Handbook of the Indians of California* (Washington, DC: Government Printing Press, 1925), 442. Trying to come by one clear name for any group of California Indians is difficult, because there is no clear set of social divisions. My own practice has been to use the most local names that I can find.

17 *Digger* eventually came to be a shorthand term for any inland California Indian, extending to those also living in Nevada and Utah. For various reasons, including their supposed extreme poverty, which forced them to eat roots, as well as their dark-brown skin color, they were among the American Indian groups most despised by white Americans. James J. Rawls, *Indians of California: The Changing Image* (Norman: University of Oklahoma Press, 1984), 32–34, 49.

18 Galen Clark, *Indians of the Yosemite Valley and Vicinity: Their History, Customs and Traditions* (San Francisco: H. S. Crocker Co., 1904), 5; and Mark David Spence, *Dispossessing the Wilderness: Indian Removal and the Making of the National Parks* (New York: Oxford University Press, 1999), 103. One of the best deep landscape histories of Yosemite, including the Mariposa War of 1851 and the landscape it helped to spawn, is Rebecca Solnit, *Savage Dreams: A Journey into the Landscape Wars of the American West* (Berkeley: University of California Press, 1999), esp. "Water, or Forgetting the Past: Yosemite National Park," 215–385.

19 Clark, *Indians of the Yosemite Valley and Vicinity*, 6–10.

20 Albert L. Hurtado, *Indian Survival on the California Frontier* (New Haven, CT: Yale University Press, 1988), 1, 13, 100–101, 211.

21 Clark, *Indians of the Yosemite Valley and Vicinity*, 12.

22 Lafayette Houghton Bunnell, *Discovery of the Yosemite and the Indian War of 1851, which Led to That Event* (Chicago: Fleming H. Revell, 1880), 54.

23 Bunnell, *Discovery of the Yosemite*, 55–56.

24 From *Daily Alta Californian*, Jan. 29, 1851, in *Exterminate Them! Written Accounts of the Murder, Rape, and Enslavement of Native American during the California Gold Rush*, ed. Clifford E. Trafzer and Joel R. Hyer (East Lansing: Michigan State University Press), 118.

25 See, for instance, ibid., 67, 75, 76: *Daily Alta California*, Feb. 4, 1853; Mar. 4, 1853; and June 6, 1853.

26 Bunnell, *Discovery of the Yosemite*, 62. Despite the Mariposa Battalion's best efforts and official pronouncement, the Yosemite Indians were never completely eradicated. See Spence, *Dispossessing the Wilderness*, 103, and esp.

chap. 8, "Yosemite Indians and the National Park Ideal, 1916–1969," 115–32; and Rebecca Solnit, "The Postmodern Old West, or The Precession of Cowboys and Indians," in *Storming the Gates of Eden: Landscapes for Politics* (Berkeley: University of California Press, 2007), 36.

27 Hutchings puts the discovery date at 1856 in *Scenes of Wonder and Curiosity in California* (New York: A. Roman and Company, 1871), 173, while D. J. Foley's *Yosemite Souvenir & Guide* (Yosemite, CA: Tourist Studio, 1902), 101, puts the year of discovery at 1857. It's unclear when the tree got the name "Grizzly Giant," although a quick search of period newspapers suggests that by 1865 the name was commonly attached to the tree.

28 Elizabeth Hutchinson, "They Might Be Giants: Galen Clark, Carleton Watkins, and the Big Tree," in *A Keener Perception: Ecocritical Studies in American Art History*, ed. Alan C. Braddock and Christoph Irmscher (Tuscaloosa: University of Alabama Press, 2009), 117; and Shirley Sargent, *Galen Clark: Yosemite Guardian* (San Francisco: Sierra Club, 1964).

29 Dilsaver and Tweed note that humans have lived in California for at least ten thousand years, but that archaeological and ethnographic data haven't yet been able to uncover traces of human residence in the Kaweah region beyond six hundred years ago. Dilsaver and Tweed, *Challenge of the Big Trees*, 15, 17. See also Julian H. Seward, *Indian Tribes of Sequoia National Park Region* (Berkeley: US Department of the Interior, National Park Service, 1935) 1, 7–10, 19, passim; Kroeber, *Handbook of Indians of California*, 479–81, 606–7.

30 In 1910 Fry and White interviewed Tharp; the transcription can be found in their *Big Trees*. See Fry and White, *Big Trees*, 10–11.

31 Dilsaver and Tweed, *Challenge of the Big Trees*, 20–24.

32 Douglas Hillman Strong, *Trees—or Timber? The Story of Sequoia and Kings Canyon National Parks* (Three Rivers, CA: Sequoia Natural History Association, n.d.) 7.

33 See Floyd L. Otter, *The Men of Mammoth Forest: A Hundred-Year History of a Sequoia Forest and Its People in Tulare County, California* (Ann Arbor, MI: Edwards Brothers, 1963).

34 Ibid., 14–16; and Eugene L. Menefee, *History of Tulare and Kings County* (Los Angeles: n.p., 1913), 20–24.

35 Hurtado, *Indian Survival on the California Frontier*, 123.

36 Dipesh Chakrabarty has eloquently written about the epistemological problem of trying to write about the irrational from within a discourse like academic history, which fetishizes rationality. See Dipesh Chakrabarty, *Provincializing Europe: Postcolonial Thought and Historical Difference* (Princeton, NJ: Princeton University Press, 2008), esp. 3–23.

37 Clark, *Indians of the Yosemite Valley*, x, xii.

38 John Muir, *Our National Parks*, in *John Muir: The Eight Wilderness Discovery Books* (Seattle: Mountaineers, 1992), 579.

39 Ludlow had a lifelong interest in psychotropic drugs and was an aficionado of hash. His *The Hasheesh Eater* (1857) was the *Doors of Perception* of its day, and widely read. Ludlow, "Seven Weeks in the Great Yo-Semite," 744.
40 Mary H. Wills, *A Winter in California* (Norristown, PA: Morgan R. Wills, 1889), 139.
41 See, for instance, Marguerite S. Shaffer, *See America First: Tourism and National Identity, 1880–1940* (Washington, DC: Smithsonian Institution Press, 2001); Kristin L. Hoganson, *Consumers' Imperium: The Global Production of American Domesticity, 1865–1920* (Chapel Hill: University of North Carolina Press, 2007); Leach, *Land of Desire*, xiii; Mary Louise Pratt, *Imperial Eyes: Travel Writing and Transculturation* (London: Routledge, 1992); and Solnit, *Savage Dreams*, 255, 263, passim. For a brief article that uncovers the flexibility of constructing one's own view using both guidebooks, photography, travel, and personal experience, see Maria Antonella Pelizzari, "Retracing the Outlines of Rome: Intertextuality and Imaginative Geographies in Nineteenth-Century Photographs," in *Picturing Place: Photography and the Geographical Imagination*, ed. Joan M. Schwartz and James R. Ryan (London: I. B. Tauris, 2003), 55–73.
42 King, *A Vacation among the Sierras*, 28–29, 33–34.
43 Lori Vermaas has a detailed reading of the multiple images that Watkins made of the Grizzly Giant, but Vermaas most concerns her analysis with the attempt to make the sequoias into a symbol of nationalism. See Vermaas, *Sequoia*, 44–55. For a provocative reading of the Grizzly Giant, Galen Clark, Carleton Watkins's photographs of each, and the multiple ways that these photos were distributed and received, see Hutchinson, "They Might Be Giants," 110–26.
44 I am drawing much of the factual information for my discussion of *Great Trees* from Lori Vermaas's chapter on Bierstadt and his painting. Vermaas, chap. 4, "The Centennial Version," 79–95. See also Schama, *Landscape and Memory*, 190–97.
45 Gilpin, *Remarks on Forest Scenery*, 1:45.
46 Ludlow, "Seven Weeks in the Great Yo-Semite," 745; Anne Farrar Hyde, *An American Vision: Far Western Landscape and National Culture, 1820–1920* (New York: New York University Press, 1990), 48–50.
47 *The Yosemite Valley, and the Mammoth Trees and Geysers of California* (New York: T. Nelson and Sons, 1870), 31.
48 King, *Vacation among the Sierras*, 32.
49 Sargent, *Seeking the Elephant*, 195.
50 King, *Vacation among the Sierras*, 35. King ended up being one of the few people to own the album of Watkins's photographs containing views of the sequoias. Hutchinson, "They Might Be Giants," 115.
51 Vermaas notes that by 1879 there were four or five thousand photographs of the Big Trees in circulation. By far the most popular scenes were the Grizzly

Giant, the various trees and logs that had been tunneled through, and what was left of Dowd's discovery tree. Vermaas, *Sequoia*, 31.
52 Hurtado, *Indian Survival on the California Frontier*, 72.
53 Kent G. Lightfoot, *Indians, Missionaries, and Merchants: The Legacy of Colonial Encounters on the California Frontiers* (Berkeley: University of California Press, 2005), esp. chap. 5, "Russian Merchants in California," 114–53.
54 Much of the following discussion on nationalism and the sequoias is influenced by Schama and Vermaas, although my approach differs in that I am interested in contested version of narrating the nation rather than the more unitary narrative that their work uncovers. See Schama, *Landscape & Nature*, 187–201; Vermaas, *Sequoia*, 3–28, passim.
55 Historians of the nation have long pointed out that the national imaginary resides upon a foundation of timelessness. See, for instance, Benedict Anderson, *Imagined Communities: Reflections on the Origin and Spread of Nationalism*, rev. ed. (London: Verso, 1991).
56 At least for a while. Since 1939 we have known it as *Sequoiadendron giganteum*. There was a host of competing names in the late nineteenth century, however, including *Taxodium washingtonianum*, *Sequoia wellingtonia*, and the one most favored by the US government and Galen Clark, *Sequoia washingtoniana*, which never really caught on. See US Department of Agriculture, *Report on the Big Trees of California* (Washington, DC: Government Printing Office, 1900), 22; Galen Clark, *The Big Trees of California: Their History and Characteristics* (Redondo Beach, CA: Reflex Publishing, 1907), 103–4; Geological Survey of California, *Yosemite Book*; Strong, *Trees . . . or Timber?*, 6; Fry and White, *Big Trees*, 105–6; Schama, *Landscape & Memory*, 187; and Peattie, *Natural History of North American Trees*, 6.
57 Geological Survey of California, *Yosemite Book*, 103.
58 I've stitched together the standard narrative of Sequoyah from "Se-Quo-Yah: The Inventor of the Cherokee Alphabet," *Christian Watchman* 18, no. 33 (Aug. 18, 1837): 132; "Description of the Cherokee Alphabet," *American Annals of Education* 2 (Apr. 1832): 181–84; Rev. F. R. Goulding, *Sal-O-Quah; or, Boy-Life among the Cherokees* (Philadelphia: Claxton, Remsen & Haffelfinger, 1870), 40–58; George E. Foster, *Se-Quo-Yah, the American Cadmus and Modern Moses* (1885; repr., New York: AMS Press, 1979); Grant Foreman, *Sequoyah* (Norman: University of Oklahoma Press, 1938); Traveller Bird, *Tell Them They Lie: The Sequoyah Myth* (Los Angeles: Westernlore Publishers, 1971); William G. McLoughlin, *Cherokee Renascence in the New Republic* (Princeton, NJ: Princeton University Press, 1986), 350–54; John Ehle, *Trail of Tears: The Rise and Fall of the Cherokee Nation* (New York: Anchor Books, 1988), 152–53, 160–61; Susan Kalter, "'America's Histories' Revisited: The Case of *Tell Them They Lie*," *American Indian Quarterly* 25, no. 3 (2001): 333–37; and Jill Lepore, *A is for American: Letters and Other Characters in the Newly United States* (New York: Alfred A. Knopf, 2002),

esp. chap. 3, "A National Alphabet," 63–90. The vast majority of contemporary histories that refer to Sequoyah can be traced to Foreman's 1938 work, which could be considered the first modern historical account of Sequoyah. This reliance, however, is problematic given that Foreman's historical evidence consists heavily of white hearsay and recollection. Kalter's article, "'America's Histories' Revisited" is the best current source that mines the metahistorical cultural politics of both white and academic privilege at the core of the standard tellings of the Sequoyah myth.

59 Foreman, *Sequoyah*, 3.
60 Sean P. Harvey has recently probed the histories of colonialism, racism, and language in *Native Tongues: Colonialism and Race from Encounter to the Reservation* (Cambridge, MA: Harvard University Press, 2015).
61 See Traveller Bird, *Tell Them They Lie*; for more on a Native literacy that existed independently of Euro-Americans, see Lisa Brooks, *The Common Pot: The Recovery of Native Space in the Northeast* (Minneapolis: University of Minnesota Press, 2008).
62 *Tell Them They Lie* has stirred a controversy amongst Native American historians, and at root is the epistemological question of how we determine the truth-value of historical sources. See, for instance, Raymond D. Fogelson, "On the Varieties of Indian History: Sequoyah and Traveller Bird," *Journal of Ethnic Studies* 2, no. 1 (1974): 105–12; and Arnold Krupat, "America's Histories," *American Literary History* 10, no. 1 (Spring 1998): 124–46.
63 My thinking on irony depends on Rainer Maria Rilke, *Letters to a Young Poet*, trans. M. D. Herter Norton (New York: W. W. Norton & Co., 1954); Hayden White, *Metahistory: The Historical Imagination in Nineteenth-Century Europe* (Baltimore: Johns Hopkins University Press, 1973); and David Foster Wallace, "E. unibus pluram: Television and U.S. Fiction," in *A Supposedly Fun Thing I'll Never Do Again: Essays and Arguments* (New York: Back Bay Books, 1997).
64 Carey McWilliams, *Factories in the Field: The Story of Migratory Farm Labor in California* (Berkeley: University of California Press, 1999), 55.
65 Rawls, *Indians of California*, xiv, 69, 81, 96–107, 141; Hurtado, *Indian Survival on the California Frontier*, 13, 76, 125, 130, 135, 147. As Carey McWilliams and Don Mitchell note, this continues to be the modus operandi of California's agricultural capitalists. See Carey McWilliams, *Factories in the Field*; and Don Mitchell, *The Lie of the Land: Migrant Workers and the California Landscape* (Minneapolis: University of Minnesota Press, 1996).
66 Jill Lepore, *The Name of War: King Phillip's War and the Origins of American Identity* (New York: Alfred A. Knopf, 1998), xi.
67 Alan Trachtenberg, *Shades of Hiawatha: Staging Indians, Making Americans, 1880–1930* (New York: Hill and Wang, 2004), esp. "Introduction: Dreaming Indian," 3–50, and chap. 4, "Ghostlier Demarcations," 170–210; Brian Dippie, *The Vanishing American: White Attitudes and U.S. Indian Policy* (Law-

rence: University Press of Kansas, 1982), 12, 67, 83, passim; Philip J. Deloria, *Playing Indian* (New Haven, CT: Yale University Press, 1998), 4, 5, 25, 26, passim; Jean M. O'Brien, *Firsting and Lasting: Writing Indians out of Existence* (Minneapolis: University of Minnesota Press, 2010); Nye, *America as Second Creation*, 2, 4, 5, 11,14, passim; Lepore, *Name of War,* esp. "Part Four: Memory," 173–226.

68 Geological Survey of California, *Yosemite Book*, 104.
69 Shepard Krech III, *The Ecological Indian: Myth and History* (New York: W. W. Norton & Co., 1999), 26–27.
70 Laurence Buell notes that "nature has been doubly otherized in modern thought. The natural environment as empirical reality has been made to subserve human interests, and one of these interests has been to make it serve as a symbolic reinforcement of the subservience of disempowered groups: nonwhites, women, and children." In other words, the natural world is ontologically separated from and of secondary importance to the human (the first otherization); this secondary world then becomes the sign of less-than-human humans (the second otherization). Buell, *Environmental Imagination*, 21.
71 Asa Gray, *Sequoia and Its History: An Address by Professor Asa Gray, President of the American Association for the Advancement of Science, Delivered at the Meeting Held at Dubuque, Iowa, August, 1872* (Salem, MA: Salem Press, 1872), 5–6.
72 Art historian Angela Miller makes a similar point, though she limits her critique to the equation of nationalism with wilderness. See Angela L. Miller, "The Fate of Wilderness in American Landscape Art: The Dilemmas of 'Nature's Nation,'" in *A Keener Perception: Ecocritical Studies in American Art History*, ed. Alan C. Braddock and Christoph Irmscher (Tuscaloosa: The University of Alabama Press, 2009), 85–109.
73 Though I am drawing throughout this section on insights gleaned from Homi K. Bhabha's articulation of mimicry (upon which Roach draws) as well as Lawrence Buell's discussion of double otherization, I have chosen to rely on Roach's related surrogation because of its flexibility in extending to the nonhuman world as well as its reliance on performance. Reading Bhabha, Buell, and Roach alongside one another can go a long way toward helping probe the grotesquerie of creating a celebratory monument to those one seeks to exterminate. See Bhabha, *Location of Culture*, esp. chap. 4, "Of Mimicry and Man: The Ambivalence of Colonial Discourse," 121–31; Buell, *Environmental Imagination*, 20–21; Joseph Roach, *Cities of the Dead: Circum-Atlantic Performance* (New York: Columbia University Press, 1996) 2. Though Roach does not directly examine an environmental aspect of surrogation, Monique Allewaert does in her analysis of literary performance and identity in the Caribbean. See Monique Allewaert, *Ariel's Ecology: Plantations, Personhood, and Colonialism in the American Tropics* (Minneapolis:

University of Minnesota Press, 2013) esp. chap. 5, "Involving the Universe in Ruins," 143–72; for a more in-depth look at Nature's Nation and Indianness, see Daegan Miller, "Reading Tree in Nature's Nation: Toward a Field Guide to Sylvan Literacy in the Nineteenth-Century United States," *American Historical Review* 121, no. 4 (Oct. 2013): 1114–40, 1129–32.

74 Krech, *Ecological Indian*, 16–17.

75 Benjamin Tompson, *New England's Crisis; or, A Brief Narrative of New England's Lamentable Estate at Present, Compar'd with the Former (but few) Years of Prosperity* (Boston: John Foster, 1676), 14.

76 Geological Survey of California, *Yosemite Book*, 108; see also Lepore, *Name of War*, 83–89.

77 For the intersecting history of the traumas of the Civil War and America's woodlands, see Aaron Sachs, *Arcadian America: The Death and Life of an Environmental Tradition* (New Haven, CT: Yale University Press, 2013), esp. chap. 4, "Stumps," 137–209; and Megan Kate Nelson, *Ruin Nation: Destruction and the American Civil War* (Athens: University of Georgia Press, 2012), esp. chap. 3, "Battle Logs: Ruined Forests," 103–59.

78 H. C. White & Co., "'The Confederates.' A Noble Group of Forest Giants, Mariposa Grove, Cal., U.S.A." (Bennington, VT: H. C. White & Co., 1903); the Robert E. Lee, in the Grant Grove, was named in 1875; in addition, a General Lee was named in 1901. See Fern Gray, *And the Giants Were Named* (Three Rivers, CA: Sequoia National History Association, n.d. [1960?]) 4, 9. John Muir, *The Mountains of California* in *The Eight Wilderness-Discovery Books* (Seattle: Mountaineers, 1992), 368.

79 Michael Fellman, *Citizen Sherman: A Life of William Tecumseh Sherman* (New York: Random House, 1995), "Indian Killer," 259–76; quotation on 264.

80 "The Modoc War," *New York Times*, Apr. 15, 1873.

81 Roach, *Cities of the Dead*, 2–6.

82 "Haunting belongs to the structure of every hegemony," writes Jacques Derrida. This is why, as Lefebvre has pointed out, space never is produced just once but is always in the process of being constantly reproduced. Jacques Derrida, *Specters of Marx: The State of the Debt, the Work of Mourning and the New International*, trans. Peggy Kamuf (New York: Routledge, 1994), 62, 64; and Henri Lefebvre, *The Production of Space*, trans. Donald Nicholson-Smith (Malden, MA: Blackwell Publishing, 1991), 69, 85, 90, 190, passim.

83 King, *Vacation among the Sierras*, 35.

84 J. M. Hutchings, *Scenes of Wonder*, 50.

85 This entire essay is an attempt to wrestle with the questions Derrida raises in *Spectres of Marx*—how best can we who live in the present live with the ghosts of the past while working toward a better future? See Derrida, *Specters of Marx*. See also Chakrabarty, *Provincializing Europe*; and Michel Rolph Trouillot, *Silencing the Past: Power and the Production of History* (Boston: Beacon Press, 1995), esp. chap. 5, "The Presence in the Past," 141–53.

86 Donald F. Durnbaugh, "Communitarian Societies in Colonial America," in *America's Communal Utopias*, ed. Donald E. Pitzer (Chapel Hill: University of North Carolina Press, 1997), 16; Carl J. Guarneri, *The Utopian Alternative: Fourierism in Nineteenth-Century America* (Ithaca, NY: Cornell University Press, 1991), 36–39; Dolores Hayden, *Seven American Utopias: The Architecture of Communitarian Socialism, 1790–1975* (Cambridge, MA: MIT Press, 1976), 261–85; the grandfather of communal studies is Arthur Bestor's *Backwoods Utopias: The Sectarian Origins and the Owenite Phase of Communitarian Socialism in America, 1663–1829* (Eugene, OR: Wipf & Stock, 1970).

87 Christopher Clark, *The Communitarian Moment: The Radical Challenge of the Northampton Association* (Ithaca, NY: Cornell University Press, 1995), 1–2; Durnbaugh, "Communitarian Societies in Colonial America," 14. Greeley may have never uttered the words we remember him for: I've been unable to find the quotation, and there's a great deal of wrangling over who, in fact, first uttered the words, and whether or not they ever came from Greeley's mouth or pen. For more on the controversy, see Robert C. Williams, *Horace Greeley: Champion of American Freedom* (New York: New York University Press, 2006), 40–43.

88 For more on George, see John L. Thomas, *Alternative America: Henry George, Edward Bellamy, Henry Demarest Lloyd and the Adversary Tradition* (Cambridge, MA: Harvard University Press, 1983), esp. chap. 5, "*Progress and Poverty*," 102–31; and Sachs, *American Arcadia*, esp. 225–36, 274–81.

89 Edward Bellamy, *Looking Backward: 2000–1887* (New York: American Library, 2000), v; Thomas, *Alternative America*, esp. chap. 10, "*Looking Backward*," 237–61.

90 Robert V. Hine, *California's Utopian Colonies* (New York: W. W. Norton & Co., 1966), 6–9.

91 Kaweah, *A Co-Operative Common Wealth: Located Above Three Rivers P.O. on the Kaweah River, Tulare Co.* (San Francisco: Book & Job Printers, 1887), 3–4.

92 Jay O'Connell, *Co-Operative Dreams* (Van Nuys, CA: Raven River Press, 1999), 59.

93 J. J. Martin to the Members of Kaweah Colony, July 5, 1887, MS, KCCP, box 1 folder 7.

94 Burnette G. Haskell, *CW* 20 (Nov. 1889): 168.

95 O'Connell, *Co-Operative Dreams*, 57. Philip Deloria notes that in the 1880s and 1890s especially, the perceived savagery and otherness of American Indians was used by many Americans as a refreshing tonic to fortify oneself against the ills of overcivilization. This certainly applies to Kaweah: one of the things socialists could admire about "authentic" Indians was their communitarian tribalness. See Deloria, *Playing Indian*, 103–6.

96 There is very little written about the Kaweah Colony. The indispensible work includes O'Connell's *Co-Operative Dreams*, which is the only book-length

treatment. Aside from sources already cited, I've also drawn from Elaine Lewinnek, "The Kaweah Co-Operative Commonwealth and the Contested Nature of Sequoia National Park," *Southern California Quarterly* 89, no. 2 (Summer 2007): 141–67; Paul Kagan, *New World Utopias: A Photographic History of the Search for Community* (New York: Penguin Books, 1975); William Tweed, *Kaweah Remembered: The Story of the Kaweah Colony and the Founding of Sequoia National Park* (Three Rivers, CA: Sequoia Natural History Association, 1986); Maria A. Reed "The Strength of Vision: A Case Study of the Kaweah Cooperative Colony" (MA thesis, University of Chicago, 2011); and Stacy Colleen Kozakavich, "The Center of Civilization: Archaeology and History of the Kaweah Co-operative Commonwealth" (PhD diss., University of California, Berkeley, 2007).

97 O'Connell, *Co-Operative Dreams*, 18–23. The First International was also known as the International Workingmen's Association, and though many have come to associate it as Marx's project, it was really a diverse group of anarchists, trade unionists, Marxists, Italian nationalists, Owenites — generally an amalgamation of any left-leaning European radical socialist. See Edmund Wilson, *To the Finland Station: A Study in the Writing and Acting of History*, foreword by Louis Menand (New York: New York Review of Books, 1972), 257–61; Hine, *California's Utopian Colonies*, 79. For Bakunin's influence in America, and, briefly, on Haskell, see Paul Avrich, *Anarchist Portraits* (Princeton, NJ: Princeton University Press, 1988), esp. chap. 2, "Bakunin and the United States," 16–31. And for a good historical context of anarchist thinking in the nineteenth-century United States, see Peter Marshall, *Demanding the Impossible: A History of Anarchism* (Oakland, CA: PM Press, 2010), esp. chap. 32, "United States," 496–503.

98 "A Brief History of Kaweah," *CW* 2, 15 (May 1889).

99 C. F. Keller to Carl Keller, Apr. 14, 1921, transcription, Kaweah Microfilm 00343 (San Marino, CA: Huntington Library Photographic Department, 1952).

100 For more on Gronlund, see P. E. Maher, "Laurence Gronlund: Contributions to American Socialism," *Western Political Quarterly* 15, no. 4 (Dec. 1962): 618–24.

101 Hurtado, *Indian Survival on the California Frontier*, 1, 100.

102 This was part of a longer historical practice in which the federal government sought to encourage landholding through the passage of various acts including the Preemption Act (1841), Homestead Act (1862), Timber Culture Act (1873), Desert Land Act (1877), and Timber and Stone Act (1878), which all offered land at cheap prices to those individuals willing to "improve" it.

103 C. F. Keller to Carl Keller, Apr. 14, 1921.

104 Burnette Haskell, "Kaweah: How and Why the Colony Died," in *Giants in the Earth: The California Redwoods*, ed. Peter Johnstone (Berkeley, CA: Heyday Books, 2001), 108.

105 J. J. Martin, "History," box 1, folder 4, 28. The title and legal standing of the organization to which the communards belonged changed a half-dozen times during its existence. Jay O'Connell does a good job of charting the changes in *Co-Operative Dreams*, and I won't get into them here.
106 Gray, *And the Giants Were Named*, 12.
107 Philip Winser, "Memories," 1931, MS, Henry E. Huntington Library, San Marino, California, 83, 92–93, 94.
108 Tweed, *Kaweah Remembered*, 2.
109 Burnette G. Haskell, *Kaweah, A Co-operative Common Wealth: Located Above Three Rivers P.O. on the Kaweah River, Tulare Co.* (San Francisco: Book & Job Printers, 1887), 2.
110 This is the most valuable insight coming from "The Trouble with Wilderness" paradigm. Ramachandra Guha put it best when he wrote, "A truly radical ecology in the American context ought to work toward a synthesis of the appropriate technology, alternate lifestyle, and peace movements." Not a bad gloss on Kaweah. Ramachandra Guha, "Radical American Environmentalism and Wilderness Preservation: A Third World Critique," in *The Great New Wilderness Debate*, ed. J. Baird Callicott and Michael P. Nelson (Athens: University of Georgia Press, 1998), 242.
111 Kaweah Colony, *Pen Picture*, 4.
112 W. Carey Jones, "The Kaweah Experiment in Co-operation," *Quarterly Journal of Economics* 6, no. 1 (Oct. 1891): 49; Kaweah Colony, *Pen Picture*, 4
113 J. J. Martin, "History," box 1, folder 5.
114 For a boosterish, contemporary take on the potential of sequoian lumber, see, for instance, Thos. H. Thompson, *Official Historical Atlas Map of Fresno County* (Tulare, CA: Thos. H. Thompson, 1891).
115 Jones, "Kaweah Experiment in Co-operation," 49, 52.
116 Haskell, "Kaweah," 106.
117 O'Connell, *Co-Operative Dreams*, 32, 37.
118 Thompson, *Historical Atlas Map of Fresno County*, 16. Road building was backbreaking work done with pick and shovel, iron bar, and dynamite charge, but by 1890, the colonists had reached the proposed mill site, and logging had begun. To give an idea of the engineering marvel that was the Kaweahans' road, it remained the only road into the Giant Forest until 1926, and it is still used today. Kagan, *New World Utopias*, 89.
119 Kaweah Colony, *Pen Picture*, back cover.
120 Burnette G. Haskell, *CW*, Nov. 1889, 123.
121 Jones, "Kaweah Experiment in Co-operation," 72.
122 J. J. Martin, "History," box 1, folder 10, 135; Hine, *California's Utopian Colonies*, 85; Lewinnek, "Kaweah Co-Operative Commonwealth," 147. For a table of occupants, and where they lived, see Kozakavich, "Center of Civilization," 446–51.
123 Laurence Gronlund, *The Co-operative Commonwealth: An Exposition in*

Modern Socialism (1891), facs. ed. of 3rd ed. (La Vergne, TN: BiblioLife, 2010), 22. American anarchist Benjamin Tucker was the first to translate *What Is Property?* for an American audience in 1876, and it was probably this 1876 edition that so influenced the Kaweahans.

124 Pierre-Joseph Proudhon and Amédée Jérôme Langlois, *What Is Property? An Inquiry into the Principle of Right and of Government*, facs. ed. (Princeton, MA: Benj. R. Tucker, 1876), 11, 42–44, 52, 88–89.

125 Proudhon, *What Is Property?*, 82.

126 Of course, Aldo Leopold is not an unproblematic environmental thinker, in terms of both his actual actions and his policies, as both Louis Warren and Donald Worster have pointed out. But I think his land ethic is worth lingering over precisely because it is so flexibly robust. Like Proudhon, Leopold's conception of ethics relies on community, although Leopold would extend the notions of rights and duties to the natural world. Both critique the culture of capitalism that seeks to extend its hegemony to everything. "All ethics so far evolved rest upon a single premise: that the individual is a member of a community of interdependent parts," Leopold writes in a Kropotkinesque vein, "the land ethic simply enlarges the boundaries of the community to include soils, waters, plants, and animals, or collectively: the land." Even more compellingly, Leopold writes, "A land ethic changes the role of *Homo sapiens* from conquerer of the land-community to plain member and citizen of it. It implies respect for his fellow-members, and also respect for the community as such." This starts to sound an awful lot like Proudhon's usufructuary. Aldo Leopold, *A Sand County Almanac with Essays on Conservation from Round River* (New York: Ballantine Books, 1966), 239, 240, 262. See also Louis S. Warren, *The Hunter's Game: Poachers and Conservationists in Twentieth-Century America* (New Haven, CT: Yale University Press, 1997), esp. chap. 3, "'Raiding Devils' and Democratic Freedoms: Indians, Ranchers, and New Mexico Wildlife," 71–125, and chap. 4, "Tourism and the Fading Forest," 106–25; and Donald Worster, *Nature's Economy: A History of Ecological Ideas*, 2nd ed. (Cambridge: Cambridge University Press, 1994), esp. chap. 13, "The Value of a Varmint," 258–90.

127 Gronlund, *Co-operative Commonwealth*, 81.

128 Gronlund, ibid., 39.

129 Donald Worster notes that in tying his land ethic too firmly to the science of ecology, which was "preparing to turn abstract, mathematical, and reductive," as well as mechanical, Leopold ended up inadvertently reinforcing the status quo. Ultimately, ecology, in seeking to scientifically, abstractly distance itself from the messy realities of the real world ended up accommodating itself quite nicely to a capitalistic ethos. See Worster, *Nature's Economy*, 289–90.

130 Fry and White, *Big Trees*, 23. The classic article on the creation of Sequoia National Park is Oscar Berland, "Giant Forest's Reservation: The Legend and the Mystery," *Sierra Club Bulletin* 47 (Dec. 1962): 68–82.

131 Dilsaver and Tweed, *Challenge of the Big Trees*, 62–64; Roderick Frazier Nash, *Wilderness and the American Mind*, 4th ed. (New Haven, CT: Yale University Press, 2001), 105–7. Of course, the intuition that water levels depended to some degree on forest cover was not invented by George Perkins Marsh; as Richard Grove has shown, the fears of desiccation had been well known since the age of exploration. Richard Grove, *Green Imperialism: Colonial Expansion, Tropical Island Edens and the Origins of Environmentalism, 1600–1860* (Cambridge: Cambridge University Press, 1995).

132 Muir, *Our National Parks*, 590–91.

133 Winser, "Memories," 103.

134 O'Connell, *Co-Operative Dreams*, 32–33.

135 For a good description of all the legal wrangling, and there was a good bit of it, see O'Connell, *Co-operative Dreams*, 32–37, 66–73, 125–28; Dilsaver and Tweed, *Challenge of the Big Trees*, 64–69, 73–83; and Berland, "Giant Forest's Reservation," 68–82.

136 Hine, *California's Utopian Colonies*, 95.

137 Richard J. Orsi, *Sunset Limited: The Southern Pacific Railroad and the Development of the American West 1850–1930* (Berkeley: University of California Press, 2005), 100–101; for a take on the railroads that is far more critical, see Richard White, *Railroaded: The Transcontinentals and the Making of Modern America* (New York: W. W. Norton & Co., 2011). J. J. Martin, "History," box 1, folder 7, 75–79.

138 The Mussel Slough affair continues to be controversial. I've drawn my account from McWilliams, *Factories in the Field*, 15–17, a telling obviously very sympathetic to the settlers, as well as Orsi's far more critical *Sunset Limited*, 102–3, 472–73n34.

139 I've stitched together Zumwalt's biographical details from Orsi, *Sunset Limited*, 75, 86, 88, 100, 194.

140 See Strong, *Trees . . . or Timber?*, 25–30; Dilsaver and Tweed, *Challenge of the Big Trees*, 69–73; O'Connell, *Co-Operative Dreams*, 190–98; Orsi, *Sunset Limited*, 363–64, 370–71; Farmer, *Trees in Paradise*, 41. Berland broke this story, in 1962. See Berland, "Giant Forest's Reservation."

141 George W. Stewart to John R. White, June 8, 1929, reprinted in Fry and White, *Big Trees*, 28.

142 Orsi, *Sunset Limited*, 363–64.

143 O'Connell, *Co-Operative Dreams*, 117.

144 Berland, "Giant Forest's Reservation," 78.

145 Anne Haskell, diary entry, Nov. 24, 1890, HFP.

146 Lewinnek, "Kaweah Co-operative Commonwealth," 158; Anne Haskell, diary entry Oct. 14, 1890, HFP.

147 O'Connell, *Co-Operative Dreams*, 126–27; Dilsaver and Tweed, *Challenge of the Big Trees*, 73–76.

148 The actual details of this long and convoluted story are these: The com-

mander of the cavalry troop, Captain Dorst, had no idea where the park began and ended. It was brand new, and no boundaries had actually been inscribed on the landscape. Given faulty information by a local land agent, and then conflicting information by the commissioner of the General Land Office, Dorst did everything but do actual physical harm to stop the colonists from logging, until he received new, clear orders from the acting secretary of the interior, George Chandler, that the colonists were, indeed, cutting on private land and that they should be left alone. See Dilsaver and Treed, *Challenge of the Big Trees*, 76–83.

149 Or at least for the time being. Kagan, *New World Utopias*, 99–100; J. J. Martin, "History," box 1, folder 8, 84.

150 J. J. Martin, "History," box 1, folders 9–10, 124, 132, passim.

151 Robert V. Hine, "California's Socialist Utopias," in *America's Communal Utopias*, ed. Donald E. Pitzer (Chapel Hill: University of North Carolina Press, 1997), 422–23.

152 Karl Marx and Friedrich Engels, *The Communist Manifesto* (New York: Bantam Books, 1992), 16; and Allen Dodworth, *Woodman Spare That Tree! Quick Step, as Played by the Dodworth Coronet Band* (New York: Firth, Pond & Co., 1848). Henry Russell also set the poem to music, probably in the late 1830s. See Henry Russell, *Woodman! Spare That Tree! A Ballad* (New York: Firth & Hall, n.d.).

153 For a history of the army in Yosemite, see Harvey Meyerson, *Nature's Army: When Soldiers Fought for Yosemite* (Lawrence: University Press of Kansas, 2001).

154 Samuel P. Hays, *Conservation and the Gospel of Efficiency: The Progressive Conservation Movement, 1890–1920* (Pittsburgh, PA: University of Pittsburgh Press, 1999), 266.

155 George Pope Morris, "Woodman, Spare That Tree!" in *Poems of George Pope Morris*, 3rd ed. (New York: Charles Scribner, 1860), 64–65.

156 Thompson, *Official Historical Atlas Map of Fresno County*, 2.

157 Strong, *Trees . . . or Timber?*, 15; Fry and White, *Big Trees*, 17–21.

158 Farmer notes that the chief dendrologist of the United States estimated that the total efficiency of sequoia logging—the percentage of wood that actually became usable lumber—was only 25 percent to 30 percent. Farmer, *Trees in Paradise*, 44.

159 Gray, *And the Giants Were Named*, 2.

160 For a far more critical perspective, see Lewinnek, "Kaweah Co-operative Commonwealth," 147.

161 David Harvey's *Spaces of Hope* is one of the most powerful academic, analytical, rigorous, and yet thoroughly, soul-stirringly utopian examples of this that I have found, and joins Murray Bookchin ("My purpose in developing social ecology," he writes, has been to offer "a radical utopian alternative—to this day, I do not eschew the use of the word *utopian*—to the present social

and environmental crisis") in his call for the Left to reclaim utopia. Anthony Giddens, too, joins his voice to this chorus in calling for a "utopian realism"; that is, a politics animated not only by what is politically feasible in the moment, but that also recognizes that feasibility itself is informed by seemingly "unrealistic" dreams given corporeal form through hard-work activism. Politics is nothing without the fantastic. See David Harvey, *Spaces of Hope* (Berkeley: University of California Press, 2000) esp. chaps. 8 and 9, "The Spaces of Utopia" and "Dialectical Utopianism," 133–196; Murray Bookchin, *The Ecology of Freedom: The Emergence and Dissolution of Hierarchy* (Oakland, CA: AK Press, 2005), 21; Anthony Giddens, *The Consequences of Modernity* (Stanford, CA: Stanford University Press, 1990), 154–58.

162 Martin, "History," box 1, folder 10, 138; Haskell, "Kaweah: How and Why the Colony Died," 120.

163 Lewis Mumford, *The Brown Decades: A Study of the Arts in America, 1865–1895* (New York: Dover Publications, 1931), 26. For more on Dickinson, care, and resistance, see Daegan Miller, "On Care in Dark Times," *Edge Effects*, Apr. 12, 2016, http://edgeeffects.net/care-dark-times/.

164 Emily Dickinson, "26," in *The Complete Poems of Emily Dickinson*, ed. Thomas H. Johnson (Boston: Little, Brown and Company), 18.

ENDURING OBLIGATIONS

1   Edward Abbey, *Down the River* (1982; repr., New York: Plume, 1991).
2   Thomas J. Campanella, *Republic of Shade: New England and the American Elm* (New Haven, CT: Yale University Press, 2003), 3, 148.
3   Donald Culross Peattie, *A Natural History of North American Trees* (1948; repr., Boston: Houghton Mifflin Company, 2007), 248.
4   Charles E. Little, one of the biographers of the United States' dying forests, writes, "We are almost certainly witnessing the accumulated consequences of some 150 years of headlong economic development and industrial expansion, with the most impressive of the impacts coming into play since the 1950s—the age of pollution." Charles E. Little, *The Dying of the Trees: The Pandemic in America's Forests* (New York: Viking, 1995), ix–x; Eric Rutkow, *American Canopy: Trees, Forests, and the Making of a Nation* (New York: Scribner, 2012), 213, 217, and more generally chap. 7, "Under Attack," 201–27.
5   Elizabeth Kolbert, *The Sixth Extinction: An Unnatural History* (New York: Henry Holt and Company, 2014).
6   It seems that Bill McKibben was yet again prophetic in his now-classic *The End of Nature*, in which he wrote that, as a result of global climate change, "the *meaning* of the wind, the sun, the rain—of nature—has already changed. Yes, the wind still blows—but no longer from some other sphere,

some inhuman place." See Bill McKibben, *The End of Nature* (New York: Random House Trade Paperbacks, 2006), 41.

7   For information on buying an elm, contact the Elm Research Institute, the sole distributors of Liberty Elms. Although the Liberty Elms seem to be the most popular, there are other strains, for sale and not patented: Riveridge Farms propagates a strain, as does the US Arboretum. Campanella, *Republic of Shade*, 173–78.

8   See "Fast Facts," http://www.wilderness.net/index.cfm?fuse=NWPS&sec=fastFacts; "Forest FAQs," http://wilderness.org/article/forest-faqs; http://www.nps.gov/aboutus/faqs.htm; Margaret Walls, "Parks and Recreation in the United States," *Resources for the Future Backgrounder*, Jan. 2009, http://www.rff.org/RFF/Documents/RFF-BCK-ORRG_State%20Parks.pdf; and "A National System," http://www.rivers.gov/national-system.php.

9   See the blog for Columbia University's Earth Institute, http://blogs.ei.columbia.edu/2012/06/04/in-your-own-backyard-mapping-communities-near-superfund-sites/, accessed June 18, 2015. One of the by-products of the nuclear industry is plutonium, which has a half-life of twenty-four thousand years. The Department of Energy has decided that storing the stuff for ten thousand years—less than half of the half-life—is good enough. Rebecca Solnit, *Savage Dreams: A Journey into the Landscape Wars of the American West* (Berkeley: University of California Press, 1999), 78–82.

10 As of this writing, the president of the Nature Conservancy is Mark Tercek, who worked at Goldman Sachs until the Great Recession of 2008, when he jumped ship for conservation. On the Nature Conservancy's accommodation to big business, see "Our Partners in Conservation," http://www.nature.org/about-us/our-partners/index.htm?intc=nature.tnav.how.test1.

11 Samuel P. Hays's *Conservation and the Gospel of Efficiency: The Progressive Conservation Movement, 1890–1920* (Pittsburgh, PA: University of Pittsburgh Press, 1999) is the classic, key work on the undemocratic mainstreaming of environmentalism in the progressive era. The newest entry into the environmental celebration of progress is the Ecomodernist Manifesto, a mixture of the classic rhetoric of capitalism with a new digitally driven, utopian neoliberalism. See http://www.ecomodernism.org/manifesto-english/.

12 Naomi Klein, *This Changes Everything: Capitalism vs. The Climate* (New York: Simon & Schuster, 2014), 195–99.

13 One of the great exceptions is Bob Marshall, the forester who helped found the Wilderness Society in the 1930s, and a man who advocated for an inhabited wilderness that would foster socialism and civil libertarianism. Marshall got his start as a teenager hiking in the Adirondacks. See Paul Sutter, *Driven Wild: How the Fight against Automobiles Launched the Modern Wilderness Movement* (Seattle: University of Washington Press, 2002), esp. chap. 6, "The Freedom of the Wilderness: Bob Marshall," 194–238.

14 Jared Farmer, *Trees in Paradise: A California History* (New York: W. W. Norton & Co., 2013), 224; Theodore Roosevelt, *The Winning of the West* (New York: G. P. Putnam's Sons, 1897), 1:1. On the cultural trope of martial manliness, authentic wilderness experience, and capitalism, see T. J. Jackson Lears, *No Place of Grace: Antimodernism and the Transformation of American Culture, 1880–1920* (Chicago: University of Chicago Press, 1981), esp. chap. 3, "The Destructive Element: Modern Commercial Society and the Martial Ideal," 97–139.

15 Murray Bookchin, *The Ecology of Freedom: The Emergence and Dissolution of Hierarchy* (Oakland, CA: AK Press, 2005), 106.

16 Walter Harding, *The Days of Henry Thoreau: A Biography* (Princeton, NJ: Princeton University Press, 1982), xi.

17 In 1900, under the heading "Withers and Dies," a contributor to the *Salt Lake Tribune* reported that the tree had been cut down "to avoid accidents." The loss didn't go unnoticed or unmourned, and some observers realized that progress had killed a very real part of a very real lived and loved landscape. A 1904 correspondent for the *Deseret Evening News* wrote with acid pen that "when it comes to iconoclasm, engineers come as near being image smashers and landmark removers as any biped that walks the earth." It turns out that in 1899 "a gentleman back in New York desired to cut down the running time between Chicago and San Francisco." And so the surveyors came to Utah, after "changing the map of Wyoming," and "stuck up little red flags, squinted through spy glasses and generally upset things," lopping off every extraneous curve, weeding out inefficiencies along the way. The writer then went on to recount a local tall tale going around that, on the first day that the track started to be rerouted, it died, "apparently preferring death to being a living lie." See "Withers and Dies," *Salt Lake (UT) Tribune*, Sept. 16, 1900, 61, 54; "One Thousand Mile Tree of Mormon Pioneer Days," *Deseret Evening News* (Salt Lake City), Mar. 19, 1904, last edition. On the felling of Witness Tree, see Fern Gray, *And the Giants Were Named: Sequoia and Kings National Parks* (Three Rivers, CA: Sequoia Natural History Association, n.d.), 12.

18 In his recent State of the Field essay, Paul Sutter argued that "the most radical point that environmental history" can make is one of decline. I admire Sutter's commitment to making history relevant for our world, though I disagree that tales of decline are themselves radical. Rather, I throw my lot in with Walter Benjamin, who wrote, "Overcoming the concept of 'progress' and overcoming the concept of 'period of decline' are two sides of the same thing." Paul Sutter, "The World with Us: The State of American Environmental History," *JAH* 100, 1 (June 2013): 119; Walter Benjamin *The Arcades Project*, ed. Howard Eiland and Kevin McLaughlin (Cambridge, MA: Harvard University Press, 1999), 460.

19 On care and environmentalism, see Val Plumwood, *Feminism and the Mas-*

tery of Nature (London: Routledge, 1993); and Daegan Miller, "On Care in Dark Times," *Edge Effects*, Apr. 12, 2016, http://edgeeffects.net/care-dark-times/.

20 Ralph Waldo Emerson, *Nature*, in *RWE*, 3. I borrow the phrase "startling unexpectedness" from Hannah Arendt, *The Human Condition*, 2nd ed. (Chicago: University of Chicago Press, 1998), 175–247, quote at 178.

21 He could have been describing today's academic job crisis. Allen Ginsberg, "Howl," in *Allen Ginsberg: Poems, 1947–1997* (New York: HarperCollins Publishers, 2006), 134.

22 Ginsberg, "America," in *Allen Ginsberg: Poems, 1947–1997* (New York: HarperCollins Publishers, 2006), 154–56.

23 Rachel Carson, *Silent Spring* (1962; repr., Boston: Houghton Mifflin Company, 1994), 297.

24 Klein, *This Changes Everything*, 23.

25 Hannah Arendt uses the phrase "sterile passivity" to evoke the dehumanization of the modern world. See Arendt, *Human Condition*, 322.

26 See John L. Thomas, *Alternative America: Henry George, Edward Bellamy, Henry Demarest Lloyd and the Adversary Tradition* (Cambridge, MA: Harvard University Press, 1983).

27 Lewis Mumford, *The Story of Utopias* (1922; repr., La Vergne, TN: Kessinger Publishing, 2010), 307; on Waldo Frank and Van Wyck Brooks, see Casey Nelson Blake, *Beloved Community: The Cultural Criticism of Randolph Bourne, Van Wyck Brooks, Waldo Frank, & Lewis Mumford* (Chapel Hill: University of North Carolina Press, 1990); and Ralph Ellison, *Invisible Man* (1952; repr., New York: Vintage Books, 1995).

28 For a wonderful book that explores social movements not usually connected to mainstream environmentalism, see Robert Gottlieb, *Forcing the Spring: The Transformation of the American Environmental Movement* (Washington, DC: Island Press, 1993).

29 This verse was not included in the version we sang in my rural, public elementary school in Reagan's America, but it was included in Guthrie's original version of the song. For a good history of the many revisions of "This Land Is Your Land," see Mark Allan Jackson, *Prophet Singer: The Voice and Vision of Woody Guthrie* (Jackson: University Press of Mississippi, 2007), esp. chap. 1, "Is This Song Your Song Anymore? Revisioning 'This Land is Your Land,'" 19–47. The version that I quote can be found on Woody Guthrie, "This Land Is Your Land," *The Asch Recordings*, vol. 1, *This Land Is Your Land* (Smithsonian Folkways, 1999), track 14. If you know of a public school that sings these anticapitalist verses, let me know!

30 Arcade Fire, "Half Light II (No Celebration)," *The Suburbs* (Merge Records, 2010).

31 E. B. White, *Here Is New York* (1949; repr., New York: Little Bookroom, 1999), 44.

32. W. S. Merwin, *Unchopping a Tree* (San Antonio, TX: Trinity University Press, 2014), 43.
33. "The genuinely modern," writes Robert Pogue Harrison, "does not chase after the new; it makes the old new again." Robert Pogue Harrison, *Dominion of the Dead* (Chicago: University of Chicago Press, 2003), 75. The conclusion that we need new, alternative foundational stories is a mainstay of the intellectual tradition to which this book belongs. See, for instance, Leo Marx, *The Machine in the Garden: Technology and the Pastoral Ideal in America* (London: Oxford University Press, 1976), 365; and David E. Nye, *America as Second Creation: Technology and Narratives of New Beginnings* (Cambridge, MA: MIT Press, 2003), 302.
34. Edward Abbey, *Desert Solitaire: A Season in the Wilderness* (New York: Ballantine Books, 1968), xii.
35. For current cancer rates, see *Cancer Facts and Figures*, 4, 14, compiled by the American Cancer Society and available at http://www.cancer.org/acs/groups/content/@epidemiologysurveilance/documents/document/acspc-026238.pdf (accessed May 3, 2012).
36. Susan Sontag, *Illness as Metaphor and AIDS and Its Metaphors* (New York: Picador, 1989), 3.
37. Sontag, in fact, explicitly writes: "The widespread current view of cancer as a disease of industrial civilization is as unsound scientifically as the right-wing fantasy of a 'world without cancer.' . . . Presently, it is . . . a cliché to say that cancer is 'environmentally' caused." But there has been far too much research to blithely dismiss the environmental cause as merely clichéd. Siddhartha Mukherjee, in his *The Emperor of All Maladies*, writes that "cancer is the quintessential product of modernity," that cancer is a cell "whose very soul has been corrupted to divide and to keep dividing with pathological, monomaniacal purpose." Mukherjee, a doctor, is no critic—in fact, he only briefly mentions the "social challenge" on page 447—and succeeds in avoiding direct provocation, although his language often implicitly suggests critique. Sontag, *Illness as Metaphor*, 70, 72, 73, 83, 84–86; and Siddhartha Mukherjee, *The Emperor of All Maladies: A Biography of Cancer* (New York: Scribner, 2010), 242, 339.
38. Carson, *Silent Spring*, 8.
39. Sandra Steingraber, *Living Downstream: A Scientist's Personal Investigation of Cancer and the Environment* (New York: Vintage Books, 1998), 60.
40. Steingraber, *Living Downstream*, 114, 131.
41. Abbey, *Desert Solitaire*, flyleaf.
42. Henry David Thoreau, "Life without Principle," in *Thoreau: Collected Essays and Poems* (New York: Library of America, 2001), 366.

INDEX

Page numbers in boldface type refer to illustrations.

Abbey, Edward: *Desert Solitaire*, 223–28; *Down the River*, 213
abolitionism, 34, 35, 54, 57, 61, 89; and socialist communitarianism, 59, 190, 264n48; and violence, 79. *See also* Alcott, Bronson; Brown, John (abolitionist); Douglass, Frederick; Garnet, Henry Highland; Garrison, William Lloyd; Hodges, Willis; Loguen, Jermain Wesley; Ray, Charles B.; Smith, Gerrit; Smith, James McCune; Thoreau, Henry David; Wright, Theodore
Across the continent, "Westward the course of empire takes its way," **147**
Adams, Ansel, 132–33, 157, 287n74
Adirondacks, 35, 63, 66, 67, 73, 74, 76; Castorland, 48; "Forever Wild," 52, 94; Great Camps, 85; Great Northern Wilderness, 50, 51, 53, 56, 66, 68, 69, 71, 77, 79, 82, 85, 86–91; industry, 88; logging, 67–68, 88; Mt. Marcy, 80, 92; and Native Americans, 49–50; park, 93, 94, 258n8; Philosopher's Camp, 47; provenance of, 49–50, 52; tourism, 81–82, 85, 86, 89; and tuberculosis, 85–86; unicorns, 48, 50; and wilderness, 48, 50, 51, 52, 53, 55, 83, 85, 86–91, 92, 93, 95; Wild Unsettled Country, 50, 56. *See also* Brown, John (abolitionist); Murray, W. H. H.; Timbuctoo
Adventures in the Wilderness, 80, 85, 97–104. *See also* Murray, W. H. H.
Agassiz, Louis, 47, 51, 65–66, 112–13
agrarianism: and anticapitalism, 57–58; aristocratic, 58, 59; eutopian, 60–61, 64–68, 73, 78, 83, 86; and Jefferson, 20–21, 58, 60
Alcott, Bronson, 34
American Anti-Slavery Society, 60
anarchy, 13, 34, 61, 190–93, 199–200, 221, 224; and abolitionism, 56–57, 59; as method, 119, 243n26, 261n36, 265n51. *See also* Kaweah Colony
Anthony's Photographic Bulletin, 120
Anti-Rent Equal Rights Party, 57
Arabian Nights' Entertainments, The, 101
Astor, John Jacob, 56

Bakunin, Mikhail, 192
Baldwin, Loammi, *Plan of the Concord River from east Sudbury & Billerica Mills*, 37–43, **38**
Barley, H. C., *1000 Mile Tree*, **159**
Barthes, Roland, 113
Bellamy, Edward, *Looking Backward*, 191
Bierstadt, Albert, *The Great Trees, Mariposa Grove*, 174, **176**, 177
Big Trees. *See* sequoia (tree)
Bird, Traveller, 181–84
Blacksville, 55, 67, 68, 75, 79, 83
Bonner, John, 2, 5. See also *Town of Boston, The*
Bookchin, Murray, 217; *Our Synthetic Environment*, 220
boundaries, 44–45; and words, 47, 52, 80–81
Brady, Matthew, 121, 156
Brook Farm, 59. *See also* socialist communitarianism
Brown, John (abolitionist): and Adirondacks, 76–81, 82, 218, 271n112; and Emerson, 54–55; and Gerrit Smith, 77; and Thoreau, 34
Brown, John (slave trader), 48
Brown, Simon, 22–23, 28, 29, 36, 37
Brown University. *See* Brown, John (slave trader)
Bruchac, Joseph, "At the End of Ridge Road," 47
Bunnell, Lafayette Houghton, 168–69, 185
Buntline, Ned, *Love at First Sight*, 107
Bunyan, John, *Pilgrim's Progress*, 149–50
Burke, Edmund, 130, 132–33, 140

Calhoun, John C., 58
capitalism: and culture industry, 118, 133, 143, 243n25, 254n80, 305n126; and dogs eating each other, 35, 79–80, 84, 167, 226; and environmentalism, 205, 309n11; and industrialization, 17–18, 42; modernity, 4, 9–11, 13, 25, 227; and production of scarcity, 84; resistance to, 25, 46, 134, 154, 193, 203, 210, 243n26; and surveying, 24–26, 30, 32. *See also* slavery
Carson, Rachel, 226; *Silent Spring*, 220
cartography, 37–38

Castorland. *See* Adirondacks
Central Pacific Railroad, 110, 135, 154
Channing, William Ellery, 16, 23, 43–44
Clark, Galen, 166–67, 169, 171–72, 173; *Indians of the Yosemite Valley and Vicinity*, 167
Clay, Cassius, 58
Clay, Henry, 54
Cole, Thomas, 89–91; and beauty, 132; *The Course of Empire*, **91**; *Essay on American Scenery*, 135; *Home in the Woods*, **54**; *Schroon Lake*, **90**
Collins, John, 60
Compromise of 1850, 54, 74, 79
Concord River: character, 16; flooding, 16, 17–18, 44; and harmony, 45; history, 16–18, 29; Lowell mills, 17. *See also* Thoreau, Henry David: and Concord River
Construction Corps of the US Army Military Railroad. *See* Haupt, Herman
countermodernity: condition of, 11, 13, 150, 209, 219–21, 243n28, 244n30, 256n108; landscapes, 13, 95, 219; and Thoreau, 28, 33, 42
Crocker, Charles, 203–4, 208
Cronon, William, 53, 69–70, 86–87
Cropsey, Jasper Francis, *Starucca Viaduct, Pennsylvania*, 142–43, **144**
Crucifixion, The, 151, **152**

daguerreotype, 107, 108, 116; and Daguerre, 114
Darwin, Charles, *The Voyage of the Beagle*, 21
Davies, Charles, *Elements of Surveying*, 22
Dickinson, Emily, 12, 212
Discovery Tree, 165
Dixon, Melvin, 70
Dodworth, Allen, "Woodman, Spare That Tree!," 206–8, **207**
Douglass, Frederick, 66, 67, 79, 87; *My Bondage and My Freedom*, 71
Dowd, Augustus T., 164–65, **166**, 179
Durant, Thomas, 110

Ellison, Ralph, *Invisible Man*, 95, 221
Emerson, Ralph Waldo, 9, 47, 81, 89; "The Adirondacs," 48–49, 50–52; countermodernity, 219; *Nature*, 4–5; "Ode," 10; *Poems*, 10; and sequoias,

169; and slavery, 53–54; and Thoreau, 23, 250n33
Emmons, Ebenezer, 49–50, 81
environmentalism, 6, 25, 45, 53, 58–59, 61, 64, 67, 69–70, 87, 93, 157–58, 168, 171, 195–96, 200–202, 208, 215–17, 220–23, 226, 305n126
Epps, Lyman, 76, 271n111
Erie Canal, 95
eutopians. *See* agrarianism: eutopian
Evans, George Henry, 35, 57–58, 62

Faux, Egbert, 121
Fitzhugh, George, 11
Follensby Pond, 48, 52, 55
Freeman's Home, 55, 68, 75, 79, 83
Frémont, John C., 31, 164, 178
Fruitlands, 59. *See also* socialist communitarianism
Fugitive Slave Law, 54, 74, 79

Gale, George, 165, 178, 180, 211
Gardner, Alexander, *Westward the Course of Empire, Laying Tracks 300 Miles West of the Missouri River*, 142, **143**
Garnet, Henry Highland, 62, 92
Garrison, William Lloyd, 56–57, 59
General Sherman (tree), 162, 171, 188, 189, 191, 194, 195. *See also* Karl Marx (tree)
George, Henry, 150, 191, 221
Giant Forest, 171, 191, 194, 195, 197, 201, 203, 210
Gilpin, William, 131, 132, 134, 177; *Remarks on Forest Scenery*, 134–35, 177
Ginsberg, Allen: "America," 219–20; *Howl*, 219
Gohlke, Frank, *Landscape, St. Paul, Minnesota*, **218**
Golgotha, 110, 112, 151–52
Gray, Asa, 169, 180, 186
Great Elm, 1–2, 178, 239n4
Great Northern Wilderness. *See* Adirondacks
grid. *See* surveying
Grizzly Giant, 169, 174, 175
Gronlund, Laurence, *The Co-Operative Commonwealth*, 193, 199–201
Guess, George. *See* Sequoyah
Guthrie, Woody, 222

Hammond, S. H., 83–85, 87; *Wild Northern Scenes*, 83–84
Haskell, Burnette, 13, 191–93, 197, 212
Haupt, Herman, 120–23, 136, 155; *Reminiscences of General Herman Haupt*, 121–23
Hawthorne, Nathaniel, "The Celestial Railroad," 149–50
Hayden, Ferdinand Vandeveer, 117–18; *Sun Pictures of Rocky Mountain Scenery*, 117
Headley, J. T., 82–83, 85, 87, 89, 91; *The Adirondack*, 82
Henderson, James H., 92
Hodges, Willis, 66–67
Holmes, Oliver Wendell, 112
Hopedale, 60
Hutchings, J. M., 165, 166, 173, 179, 189

irony, 42, 185, 190, 218, 289n92

Jacksonian Democrats, 34
Jefferson, Thomas: and agrarianism, 20–21, 58, 60; and racism, 64–65; and surveying, 19–20, 24, 25, 32
Jemison, Edwin Francis, **100**
Jones, William, 68

Karl Marx (tree), 195, **196**, 206. *See also* General Sherman (tree)
Kaweah, and American Indians, 170–71, 192
Kaweah Colony: beginnings, 192–95; demographics, 209–10; endings, 201–6, 218, 306n148; environmental ethics, 196–97, 199–201; logging, 195–98; naming of, 192; and timecheck system, 198–99. *See also* Crocker, Charles; Gronlund, Laurence, *The Co-Operative Commonwealth*; Haskell, Burnette; Keller, C. F.; Martin, J. J.; Proudhon, Pierre-Joseph; Southern Pacific Railroad; Stewart, George; Winser, Philip; Zumwalt, Daniel K.
Keene Valley, 92
Keller, C. F., 194
King, Thomas Starr, 167, 173, 177–78, 189
Klein, Naomi, *This Changes Everything*, 220

316 INDEX

landscape: alternative, 12–14, 36, 37, 42, 44, 52, 60, 66, 106, 140, 147–48, 151, 195–96, 200, 208, 210–11, 219; as critique, 9, 11, 12, 33, 35, 123, 125, 133, 141–43, 221–23; definitions, 5–6, 25, 47, 52, 64, 88, 129–32, 134, 169, 170, 171, 177, 214, 260n29, 275n154; and identity, 2, 4, 5–6, 7, 18, 24, 29, 38, 45, 48, 58–59, 71, 92, 95, 97, 132, 168, 217; as method, 6, 25, 87, 90–91, 173, 221. *See also* wilderness
Leopold, Aldo, 200, 201, 305n126
Lessons from an Apple Tree, 6–7, 10
Leutz, Emanuel: *Washington Crossing the Delaware*, 121, **122**; *Westward the Course of Empire Takes Its Way*, **101**, 121
Loguen, Jermain Wesley, 62–63, 95
Lowell, James Russell, 47
Ludlow, Fitz-Hugh, 172, 173, 174, 177

MacLeish, Archibald, *Land of the Free*, 15
Manifest Destiny, 2, 7, 12, 32, 35, 63, 67, 118, 141–42, 149, 155, 157, 185, 187–88. *See also* O'Sullivan, John L., and Manifest Destiny
Mariposa Grove, 169, 173, 201, 204
Marsh, George Perkins, *Man and Nature*, 95, 201
Martin, J. J., 192, 193, 197, 199, 203–4, 206
Marx, Karl, 13, 164, 192, 193, 206, 224, 239n5, 243n26
Mason, David, *Ludlow*, 161
Maw, Thomas. *See* Sequoyah
Melrose, Andrew, *Westward the Star of Empire Takes Its Way—Near Council Bluffs, Iowa*, 141–42, **142**
Melville, Herman, *Moby Dick*, 21, 71
Merwin, W. S., *Unchopping a Tree*, 223
Mexican-American War, 8, 31, 34, 45, 66
modernity, 13, 221; alienation, 8–9, 13; and antimodernity, 10–11; environmentalist, 215; improvement, 4, 17; nature, 8–9; technology, 3, 9, 115, 141, 215, 217; and Thoreau, 39, 42; time, 8; towards a definition of, 8–10, 242n21, 243n25, 284n45. *See also* capitalism; countermodernity; progress
Moriah Central School, 92
Morris, George Pope, "Woodman, Spare That Tree!," 206, 207–9

Morton, Samuel George, 65; *Crania Americana*, 65
Mother of the Forest, 165, 166
Muir, John, 70, 156, 166, 169, 171, 172, 173, 201, 202, 208
Murray, W. H. H., 80–81, 85, 87, 97. *See also* "Murray's Fools"
"Murray's Fools," 97. *See also* Murray, W. H. H.
mutual aid, 60, 64, 78, 200, 201, 209, 221

Nature's Nation, 5, 187
Newberry, J. S., 118
New Harmony, 59. *See also* socialist communitarianism
Northampton Community, 59. *See also* socialist communitarianism
North Elba, NY, 55, 73, 74, 77, 78, 80
Noyes, John Humphrey, 60

Oneida Community, 60
O'Sullivan, John L., and Manifest Destiny, 149, 241n17

Pacific Railway Act, 117, 135
Perham, B. F., 37
Plumbe, John, Jr.: and daguerreotypy, 115–16, 120, 281n33; *Memorial against Mr. Asa Whitney's Railroad Scheme*, 116; and railroad dreams, 114–15, 116–17, 135; *Sketches of Iowa and Wisconsin*, 114–15; suicide, 117
Poe, Edgar Allan, 113
polygenism, 65, 66. *See also* Agassiz, Louis
Price, Uvedale, 131, 132
Prine, John, "Paradise," 1
progress, 2–5, 8, 9, 10, 11, 12, 13, 17, 48, 84, 134, 141, 217, 227; and alternatives, 6, 12–13, 30, 33, 71, 140, 149, 154–57; and anxiety, 5; and technology, 51, 84, 106–7, 140, 141–42, 155–56, 162, 218. *See also* modernity
Proudhon, Pierre-Joseph, 192, 199–201; *What Is Property?*, 199–200
Proust, Marcel, *Within a Budding Grove*, 97
Pynchon, Thomas, *Against the Day*, 105

race, definition, 64
Ray, Charles B., 62–63, 73
Raymond, H. J., 82
Republican Party, 34

Reynolds, J. N., *Mocha Dick*, 21
Richardson, Albert D., *Beyond the Mississippi*, 162–63, **163**
Roche, Thomas, 126–27
Roosevelt, Theodore, 216–17
Russell, A. J., 13, 221; and artistic training, 106; and Civil War, 120–27, 135; *Coal Beds of Bear River*, 144, **145**; *Coalville, Weber Valley*, 144–45, **146**, 150; *Dial Rock, Red Buttes, Laramie Plains*, **148–49**; *East and West Shaking Hands at Laying Last Rail*, 110–12, **111**; *East Temple Street, Salt Lake City*, 150–51, **151**; *Engine "Government" down the "Banks" near Brandy, April 1864*, 124–25, **125**, 141; *Gen'l H. Haupt*, **122**; *The Great West Illustrated*, 136, 139, 143, 145–47, 289n89; *Hall's Fill above Granite Canon*, **104**; *Hanging Rock, Foot of Echo Canon*, 139–40, **140**; *Malloy's Cut*, 137–38, **137**, 139; new topographics, 293n126; *1000 Mile Tree*, 109–10, **109**, 117, 118, 119, 125, 141, 147, 152, 158, 218; *On the Mountains of Green River*, 138–39, **138**; and photographic procedure, 105–6, 107–9, 129; *Promontory Trestle Works*, 154, **155**; *Ruins in Richmond, VA*, **128**; sketch of Dutch Gap Canal, 126–27, **127**; *Skull Rock*, 152, **153**; *Stone wall, rear of Fredericksburg with rebel dead*, **124**, 141; and Union Pacific, 110, 136–41, 143–49, 150–59; and visual aesthetics, 133–34, 140–41, 147–48, 156, 157–59. *See also* Roche, Thomas; Smith, O. C.

Savage, James D., 167–69
Sedgwick, Stephen J., 118; *Illuminated Lectures across the Continent*, 118
sequoia (tree): and cultural anxiety, 189–90; and cultural nationalism, 179–80, 184, 185–89; cultural reception, 171–78; discovery by Euro-Americans, 164–66; naming controversy, 180; natural history, 161, 163–64, 166; tourism, 178–79. *See also* Bunnell, Lafayette Houghton; Clark, Galen; Discovery Tree; Dowd, Augustus T.; Gale, George; General Sherman (tree); Giant Forest; Gray, Asa; Hutchings, J. M.; Karl Marx (tree); King, Thomas Starr; Ludlow, Fitz-Hugh; Mariposa Grove; Mother of the Forest; Savage, James D.; Sequoyah; Sherman, William Tecumseh; Wills, Mary H.
Sequoia National Park, 162, 164, 166, 202, 205, 208, 210–11. *See also* Kaweah Colony; Stewart, George
Sequoyah, 180–84, **183**
Sherman, William Tecumseh, 188–89
Sims, Thomas, 54
sixth extinction, 214–15
slavery: and agrarianism, 58–61; and capitalism, 32; and Emerson, 54–55; Mexican American War, 8, 32; and Gerrit Smith, 56–58; and Thoreau, 34–36; and transcendentalists, 34; and violence, 79–80
Smith, Gerrit, 35, 77; and Adirondacks, 55–56, 57, 62, 63, 66–67, 68–69, 72, 75, 206; and John Brown, 77; and eutopian agrarianism, 60–61; and Garrison, 56–57; and slavery, 56–58. *See also* Evans, George Henry
Smith, H. Perry, *The Modern Babes in the Woods*, 87
Smith, James McCune, 13, 62, 63, 64, 66; and Adirondacks, 72–75, 87, 89, 95; biography, 71–72; on John Brown, 78; and violence, 79
Smith, O. C., 119
Smith, Peter, 55–56
socialist communitarianism, 35, 59–61, 190–91
Sogwili. *See* Sequoyah
Sontag, Susan, *Illness as Metaphor*, 225, 226
Southern Pacific Railroad, 197–98, 202–5, 211. *See also* Crocker, Charles
Stanford, Leland, 110
Stewart, George, 201, 208, 211
Stillman, W. J., 47, 82, 91
Stoddard, Seneca Ray: *The Adirondacks Illustrated*, 88; *John Brown's Grave and the "Big Rock," North Elba*, **81**
Street, Alfred B., 87
surrogation, 187–88, 189, 300n73
surveying, 18–21, 33; and abstraction, 24–26, 37; and the devil, 36; and exploration, 21; and the grid, 19–20, 24, 26, 31, 32; Jefferson, 19–20, 24, 25, 32; and literature, 21; Philadelphia, 19; and violence, 31. *See also* Thoreau, Henry David: surveying

Sutter's Mill, 164
Sweet, Homer, *Twilight Hours in the Adirondacks*, 87

Tait, Arthur Fitzwilliam, *A Good Time Coming*, **98**
Talbot, Charles. *See* Concord River: history
Texas, 32. *See also* Mexican-American War
Tharp, Hale, 169–71, 172, 191
Thomas, John, 75
Thoreau, Henry David, 6, 13, 195, 215, 217, 221; Adirondacks, 48, 91; and John Brown, 34; "Civil Disobedience," 34, 206; and Concord River, 16, 26, 29–31, 33, 35, 37–43, 44, 45, 52, 252n54; "Huckleberries," 43; jail time, 33–34; and Martin Luther King Jr., 34; "The Last Days of John Brown," 34; "Life without Principle," 33; and love, 26, 36, 228; and maps, 37–38; "Martyrdom of John Brown," 34; "Plan of Concord River," 39–43, **40**, 44; "A Plea for Captain John Brown," 34; progress, 33; railroad, 150; slavery, 34–36; surveying, 16, 21–24, 26–31, 35, 36–43, 45; and trees, 16, 26; *Walden*, 6, 33, 36, 45, 70; Walden Pond, 35, 37, 91; "Walking," 22, 35–36, 44; *A Week on the Concord and Merrimack Rivers*, 29, 30, 34, 43; wilderness, 35, 37; wildness, 35–36
Thousand-Mile Tree, **109**, 110, 112, 159, 218, 310n17
Timbuctoo, 13, 55, 68, 73, 75, 79, 83, 86, 92, 195, 260n31
time zones, 113
Tocqueville, Alexis de, 3, 4
Town of Boston, The, 2, **3**, 5. *See also* Bonner, John
transcontinental railroad, 13, 98. *See also* Plumbe, John, Jr.
Trudeau, Edward Livingston. *See* Adirondacks: and tuberculosis
Tule Indian War. *See* Kaweah, and American Indians

unicorns. *See* Adirondacks
Union Pacific Railroad, 109, 110, 123, 135, 141, 142, 146, 154
US Army Photographic Corps. *See* Russell, A. J.: and Civil War

visual aesthetics, 129–33, 141–43
visual ecology, 120, 129, 284n45, 291n111

Watkins, Carleton, *Grizzly Giant*, 174, **175**
Weber Canyon, UT, 109, 110, 119
White, Richard, 61
Whitman, Walt, 115–17
Whitney, J. D., 186
wilderness, 3, 4, 50, 51, 52–53, 62, 64, 67, 82, 83, 85, 86–91, 92, 95, 169, 170, 196, 215, 218, 274n147, 274n150, 275n153, 275n155, 292n125, 304n110; as home, 61, 71; and race, 69–71, 269n87, 269nn92–93; and trees, 87–91; Wilderness Act, 53; and work, 61, 64, 71. *See also* Adirondacks; agrarianism: eutopian
wilderness cure. *See* Adirondacks: and tuberculosis
wildness, 43, 44, 46. *See also* Thoreau, Henry David: wildness
Wild Unsettled Country. *See* Adirondacks
Wills, Mary H., 173
Winser, Philip, 195
witness tree: definition, 2, 19, 44; and identity, 5–6, 12, 13, 83, 185, 217, 228. *See Also* Russell, A. J.
Wright, Theodore, 62–63

Yosemite Valley, 162, 166, 167–69; and American Indians, 167–69; and naming, 169; and National Park, 164, 201, 204, 208, 215. *See also* Bunnell, Lafayette Houghton; Grizzly Giant; Mariposa Grove; Tharp, Hale

Zealey, J. T., 112–13
Zumwalt, Daniel K., 203, 204–5, 208, 211. *See also* Southern Pacific Railroad